SIMON FRASER UNIVERSITY
W.A.C. BENNETT LIBRARY

ORGANIC MECHANOCHEMISTRY AND ITS PRACTICAL APPLICATIONS

ORGANIC MECHANOCHEMISTRY AND ITS PRACTICAL APPLICATIONS

Zory Vlad Todres

Taylor & Francis
Taylor & Francis Group

Boca Raton   London   New York

A CRC title, part of the Taylor & Francis imprint, a member of the
Taylor & Francis Group, the academic division of T&F Informa plc.

Published in 2006 by
CRC Press
Taylor & Francis Group
6000 Broken Sound Parkway NW, Suite 300
Boca Raton, FL 33487-2742

International Standard Book Number-10: 0-8493-4078-0 (Hardcover)
International Standard Book Number-13: 978-0-8493-4078-9 (Hardcover)
Library of Congress Card Number 2005054905

Library of Congress Cataloging-in-Publication Data

Todres, Zory V., 1933-
    Organic mechanochemistry and its practical applications / Zory Vlad Todres.
        p. cm.
    Includes bibliographical references and index.
    ISBN 0-8493-4078-0 (alk. paper)
    1. Mechanical chemistry. 2. Chemistry, Organic. I. Title.

QD850.T63 2006
547--dc22   2005054905

Taylor & Francis Group
is the Academic Division of Informa plc.

Visit the Taylor & Francis Web site at
http://www.taylorandfrancis.com

and the CRC Press Web site at
http://www.crcpress.com

Preface

The principal aim of this book is to correlate mechanical actions on organic substances with the molecular events caused by such actions. Organic mechanochemistry considers conversion of mechanical energy into the driving force for molecular or structural phase transitions.

Mechanochemistry of inorganic materials is a well-developed part of the science. Chemical engineering has necessitated that inorganic mechanochemistry be addressed first. In particular, mechanical treatment was widely used to enhance the activity of metal and oxide catalysts. Regarding organic mechanochemistry, it was in its infancy for a long time. However, with increased need for high-tech applications, this part of the science has overgrown the state of latent extension and entered the period of concept formation and exploitation.

In orderly fashion, this book presents odd data on skeletal, conformational transformations, and structural phase transitions as a result of mechanical activation. All these changes take place on cutting, crushing, kneading, drilling, grinding, friction, lubrication, shearing, and sliding of organic compounds or mixtures containing these compounds. Mechanochemically induced spectral changes (so-called mechanochromism) are also considered. New data on the relationship between organic tribochemistry and organic chemistry of high pressures and shock waves are discussed.

When relevant, likely or already-realized technical applications are highlighted throughout the chapters. At the same time, obsolete concepts are replaced by new theories for boundary lubrication, with the participation of organic additives; for the polymerization of monomers, organic reactivity; and so on.

Naturally, such approaches require examination of the latest literature sources. However, the initial publications also are included when needed to complete the picture. The author index demonstrates the connections of old and modern contributions to organic mechanochemistry and, especially, the interrelation among data resulting from efforts of different scientists and inventors around the world. One part of the book considers application and behavior of organic compounds as constituents of biomechanical formulations.

Knowledge of molecular principles of organic mechanochemistry is crucial in the search for new materials, compositions, and additives that work well. This description of the scope of the book reveals the audience, which encompasses all the mechanochemistry practitioners, mechanical engineers, and organic chemists in general, including advanced students. As the mother science, organic chemistry is changing with technical progress, and its contemporary body is crossed by neighboring scientific disciplines. I hope that this book helps those young and mature specialists who wish to enter the field or those working in related fields who wish to become up-to-date on current advances. This may not be a book for faint-hearted undergraduates, but it can turn out to be a guide to students who are about to enter the job market.

Author

Zory Vlad Todres earned a doctor of philosophy degree in industrial organic chemistry from the Moscow (Russia) Institute of Fine Chemical Technology and a doctor of science degree in physical organic chemistry from the Russian Academy of Sciences. At present, he is serving as a science analyst. In the former Soviet Union, he was a leading scientist at the Institute of Organo-Element Compounds and a professor at Oil Technical University. He taught organic mechanochemistry. Within that specialty, he worked as a senior researcher for oil-lubricant manufacturing companies in Cleveland, Ohio (U.S.A.). Todres has been recognized by the international scientific community, receiving invitations as a lecturer at conferences and universities in Sweden, Portugal, Italy, Denmark, Israel, and Germany. Professor Todres authored many articles, reviews, books, and patents. His latest book, *Organic Ion Radicals — Chemistry and Applications*, was published in 2002 by Marcel Dekker Publishers. The World Academy of Letters honored him as a member at the Einsteinian Chair of Science. He is also cited in the U.S. *Who's Who* series, particularly in *American Outstanding Professionals*.

Contents

1 Specificity of Organic Reactivity on Mechanical Activation

1.1 INTRODUCTION

Being stretched, partially neutralized polyacid fibers expel protons in the surrounding bath, which results in a measurable drop of pH (Bisio et al. 2003). This mysterious happening is explained below. What is fundamental is that this example introduces systems that are able to transform mechanical action into a chemical driving force.

From the end of the 19th century, it has been known that some chemical substances react differently when exposed to mechanical and thermal energy. At that time, works by Carey Lea introduced mechanochemistry as a separate branch of chemistry (Takacs 2004). The term *mechanochemistry* was proposed by Ostwald in 1891 for the corresponding branch of physical chemistry. In this sense, mechanochemistry should be considered along with thermochemistry, electrochemistry, photochemistry, sonochemistry, chemistry of high pressures, shock waves, or microwave effects.

This book discusses chemical reactivity of organic molecules mechanically treated separately or cojointly. First, mechanical treatment enhances structural irregularity of a solid substance. Density of the lattice imperfection is substantially increased when a solid is subjected to mechanical stress during its handling by attrition, indentation, friction, comminution, compression, crushing, fracturing, kneading, swelling, rubbing, milling, or grinding.

In fact, organic solids are very liable to attrition; inorganic materials appear to be more resistant (Bravi et al. 2003). Attrition represents wear caused by rubbing or friction. When an object moves along a surface or through a viscous liquid or gas, the forces that oppose its motion are referred to as *friction*. Frictional forces are nonconservative, converting kinetic energy of the slide-contacting materials into their internal energy. Friction and other kinds of mechanical action lead to an increase in enthalpy and, sometimes, result in the formation of thermodynamically metastable states. Such increase affects an equilibrium state and the kinetics of a reaction between the mechanically activated molecules. In some cases, organic reagents are activated separately and then mixed to launch the reaction between them. When they are mechanically treated cojointly, organic substances interact immediately. Chemical forces arise from summation of excess enthalpy of individual participants and that of chemical interaction.

The mixing of solids normally increases the mutual solid surface area, namely, the boundary area of solids. This provides the same effect as liquid mixing. The mechanically forced reaction starts at the reagent-substrate interface. The mechanical force also improves the diffusivity of chemical species through the solid. This is one of the important factors in controlling the yield of the mechanically induced reactions.

Although the basic principles involved in transformation of mechanical energy in chemical driving force have been the subject of several reviews (e.g., Hsu et al. 2002; Kajdas 2001; Kajdas et al. 2002), some essential aspects and the most frequently discussed manifestations are presented to form a basis for the subsequent chapters dealing with particular processes.

1.2   SUBATOMIC RESULTS OF MECHANICAL ACTIVATION

Moving relatively to each other, metal surfaces emit electrons into an organic material placed between them as proved many times. A device was described to count the number of electrons emitted from copper and iron metals during rubbing (Momose and Iwashita 2004). This electron emission phenomenon was observed only while metal surfaces were mechanically rubbed with a piece of something [poly(tetrafluoroethylene) in the authors' case] between them. This piece acts as a sponge for electrons emitted.

When two surfaces slip each other, wear-free friction can primarily occur because of the vibration of the atomic lattices. The atoms close to one surface vibrate when the outer atoms in the opposing surface slip across them. These vibrations are called *phonons* or sound waves. (A phonon is a quantum of sound energy that is a carrier of heat.) The phonons dissipate energy as heat, and this microscopic process is conjugated with friction. Phonon dissipative mechanisms involve the direct transfer of energy into the phonon populations. In contrast, electronic mechanisms involve energy transfer into conduction electrons before transfer to the phonon populations.

As follows from molecular dynamics calculations, in 10^{-10} sec after application of mechanical stress, the energy balance of the macromolecule destruction is: one third of the total energy transforms into the phonon energy, one third is expended for the high-energy vibration of excited states, and one third is for the fragmental destruction of the polymer (Zarkhin and Burshtein 1982).

During shear deformation, the dimer of arylindandione undergoes splitting with the formation of radicals according to Scheme 1.1. Over 90% of the stress energy is lost as heat; that contributing to the bond cleavage is only about 1% (Dadali et al. 1992). The breaking bond is weak, and its splitting energy is 75 kJ/mol (Nikulin and Pisarenko 1985). As is known, pressure contracts reaction volumes and promotes many organic reactions. Nevertheless, pressure alone (up to 0.7 GPa), without shear strain, does not initiate homolysis of bis(arylindandione). It is the shearing component of the mechanical stress that causes the bond rupture according to Scheme 1.1 (Dadali et al. 1988).

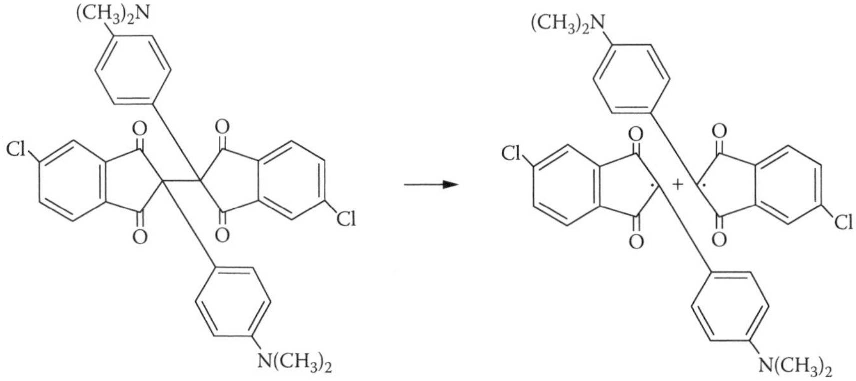

SCHEME 1.1

There are many other kinds of energy generated upon friction. Kajdas (2005) gives a ramified scheme including radio waves, light and acoustic emission, and electric and magnetic field effects. Certainly, everything affects everything, but we consider only those effects that are now known as having clear chemical manifestation.

1.3   GENERAL GROUNDS OF MECHANICALLY INDUCED ORGANIC REACTIONS

To add to the consideration of the pressure effect, let us direct our attention to pressure-induced light emitting from organic luminophores. Chapter 2 gives a number of examples of pressure enhancing light emission from photoexcited organic compounds. Here, however, one case is considered of pressure weakening luminescence intensity. An exception sometimes helps improve understanding of the regularity. Photoexcitation of 6-aminocouimarin induces intramolecular electron transfer, which in its utmost form, is represented by Scheme 1.2. The amino group can rotate, giving rise to the twisted conformation. In respect to nitrogen of the amino group, sp^2 planar hybridization can be changed by sp^3 pyramidal hybridization one on excitation. In any case, two excited states, planar and twisted, are formed. As shown, just the twisted state is responsible for luminescence of 6-aminocoumarin. Pressure leads to the loss in intensity of the emission. The loss with increasing pressure is explained two ways. A decrease in efficiency can be caused by an increased rate of

SCHEME 1.2

nonradiative energy dissipation. The other reason consists of a decrease in energy transfer caused by an increase in the energy barrier between the two excited states. As a result, the light emission decreases when the pressure rises (Dadali et al. 1994 and references therein).

To explain the above-mentioned stretching effect on pH drop of the medium surrounding partially neutralized polyacid fibers, let us consider states of the carboxylic function connected with the polymer backbone. In the unstressed state, the fibers become entangled so that the carboxylic groups of different fibers draw together. The neutralized function, say COO⁻Na⁺, interacts with the nonneutralized groups because of coordinative bonding, for example, in the way of COO⁻Na⁺···HOOC or even COO⁻Na⁺···2(HOOC). Such coordination and steric hindrance factors prevent the proton elimination and hydration (to form H_3O^+) from leaving the tangle. However, stretching unbends the fiber, so that the coordinative bonds are broken, steric obstacles disappear, nonneutralized carboxylic functions dissociate freely, proton hydration proceeds without obstacles, and the surrounding water is acidified.

1.4 RELATIONS BETWEEN ORGANIC MATERIAL PROPERTIES AND MECHANICAL EFFECTS

Model instruments permit observation of chemical reactions under control of forced mechanical migration of individual reacting species (atoms, molecules, free radicals). The work of species transfer is done by molecular mills and manipulators such as an atomic force microscope, scanning tunneling microscope, and other devices. Using these instruments, atoms or molecules can be placed (positioned) into a definite position with respect to each other (with allowance for the distance between them and orientation needed for the reaction to be initiated). For example, the atomic force microscope permits the surfaces to be positioned in relation to one another with an accuracy of ~0.01 nm on action of 0.01 nN compression force. The capacities and prospects for the development of the "positional" mechanochemical reactions were described in detail by Drexler (1992). Of course, local steric and electronic effects play a certain role in positional mechanosynthesis; however, mechanical positioning has the crucial influence on the reaction rate.

Chemical properties of gases and liquids are to a large extent defined by their molecular nature. When the same atoms or molecules form solids, the cooperative interactions may make the properties of solids noticeably different from those of the individual species. The mecahnochemical reactions of solids during grinding are often beyond the scope of equilibrium thermodynamics, mainly because of the existence of short-lived, extremely activated local sites (Heinicke 1984).

Like many solid-state chemical processes, mechanochemical reactions do not proceed in the entire bulk of a solid or at the whole surface, but at certain points. These points are usually the contacts between the particles or tips of moving cracks. They are the loci of a stress field, shear deformation, and the emergence of local high temperatures and pressures. The defects in structure can serve as potential centers at which the reaction starts. So, definite localization of the chemical process takes place. Defects are also important for the transport of mechanical disturbances in solids, such as crack development, diffusion of ions, and transfer of electrons through a layer of the product formed. A solid-state reaction is essentially the reaction at the interface between the starting reactant and the product. This is the reaction zone for the processes mechanically activated.

An interface is in continuous interaction with its environment. The interaction leads to continuous changes in local composition and local surface structure (Gutmann and Resch 1996). By mechanical force action, the defect amount on the surface increases. Point defects are known to migrate within the lattice by changing place with other parts of rubbing surfaces. Because of the motion of the point defects, so-called vacancy lattice is established. Energy is preferably stored by it.

Of course, mechanochemical activation is related to changes in the interface area and lowers the phase transition temperature. For instance, crystalline zinc sulfide is transformed into wurtzite at about 970°C, but a vibromilled sample was found to undergo the phase transition at 750°C (Imamura and Senna 1982). This lowering of phase transition temperature shows that the milled substance is well developed to reach the maximal values of heat capacity and entropy changes that result in the unexpectedly high chemical reactivity.

Mechanical disintegration leads to an increase in the surface area of solids. However, this is considered a minor factor, which contributes only to 10% of the reactivity increase. The more important effect is caused by the accumulation of energy in lattice defects. Stress always presents in the lattice, and it is intensified on mechanical action. The energy accumulated in lattice defects can relax either physically by emission of heat or chemically by the ejection of atoms or electrons, formation of excited states on the surface, bond breakage, and other chemical transformations. Sometimes, milling provokes self-propagating or even explosivelike behaviors. One example is the mechanochemical self-propagating reaction between hexachlorobenzene and calcium hydride. The reaction leads to the formation of benzene and calcium chloride (Mulas et al. 1997).

Several concurring factors have been advanced to explain the self-propagating character of this dechlorination. Local activated states develop in the mechanically processed powder because of accumulation of structural defects, vacancies, dislocations, and intergranular boundaries. These factors promote diffusive events, chemical interactions, and spontaneous structural transformations. The continuous cleaning of the available surfaces by the milling action limits diffusion influence and prevents the products from retardation of further reaction. Higher intensive grinding acts to increase both the defect content and the accessible reaction area by further reduction of the powder particle size. More activated sites are formed, and more excess energy is accumulated in the extended network of intergranular boundaries. Such an extended network becomes available for a chemical reaction.

Mechanically induced generation of organic free radicals is one of the most illustrative examples of a mechanically induced reaction. Chapter 4 discusses a series of corresponding reactions. Here we discuss the surface events that assist the radical recombination. Let us return to Scheme 1.1, which depicts the free-radical generation. The dimer in the prerupture state is stretched, and after bond cleavage, the radicals are removed from each other and leave the unit volume. The rupture of the junction bond is the limiting step of the reaction. The emergence of the radicals from the cage of incipiency is the nonlimiting step.

When they have emerged from the cage, these free radicals undergo recombination and decomposition. In addition, mechanical treatment causes plastic deformation of the material and migration of the particles formed. In diffusing, the radicals encounter each other and can be coupled and substituted.

The reactions occurring in the field of mechanical stress should be distinguished from those proceeding after the field has been removed (the posteffect). The former type of reactions would most likely involve mechanically activated species. In the latter type of reaction, the role of mechanical action reduces mostly to the generation of radicals, while the subsequent free-radical process obeys the rules inherent in them.

The sense of mechanochemical phenomena dictates reformulation of some principles from physical chemistry. For example, activation energies of thermooxidation of polypropylene and polyethylene are reduced in conditions of tensile stress (Rapoport and Zaikov 1983). According to the authors, tensile stress evokes a strain of those bonds in the polymers that would be oxidized. Speaking in terms of le Chatelier's principle, we should say the following: When the active site is elongated in the course of a reaction, tensile stress increases, and pressure decreases the rate. Reversibly, when the reactive site is shortened, tensile stress decreases, and pressure increases the rate. Energy enrichment of binding sites that takes place during mechanical strain also leads to an increase in their proton affinity (Beyer 2003).

Two mechanisms of mechanochemical reactions are most likely. First, under the action of mechanochemical stress, intermixing at the molecular level occurs. Second, the product is formed on the surface of macroscopic reacting species. In the special case of explosion reactions, they are initiated by a shock wave, which loosens the lattice of reacting particles for the period of the relief of elastic stress and thus make the system quasi-homogeneous.

When organic fluids are confined in a narrow gap between two smooth surfaces, their dynamic properties are completely different from those of the bulk. The molecular motions are highly restricted, and the system shows "solidlike" responses when sheared slowly. Solidification leads to layering of molecules at surfaces and to a decrease in molecular volumes because of confinement. These two effects appear simultaneously, but the extent of their contribution could be different for various confined "solidified" systems.

Mechanical treatment of solids initiates chemical processes. These processes are caused by a series of phenomena developed on the treatment. Zones of local warming occur at contacts between solid particles or inside these particles. High pressure and shear stress also emerge in the contact places. Mechanical treatment of solids eventually generates heat. Such thermal effect can be sufficiently strong, leading to local temperature pulses of up to 1000 K. If mechanical energy is provided faster than heat evolution takes place, physicochemical transformations of the substrate occur at the molecular level. These transformations include vibration and electronic excitation with rupture and formation of valence bonds and deformation of bond angles. On surfaces, the energy-rich domains originate. These hot spots initiate chemical reactions. Electron transfer processes (redox reactions) and ionization take place on mechanical action. The tribochemical formation of ions and radicals belong to the triboelectrical events (Sakaguchi and Kashiwabara 1992).

On the other hand, both destruction and generation of weaker intermolecular interactions (disordering, amorphization of the crystal structure, conformational transitions, and polymorphic transformations) can occur under mechanical activation. Regarding organic substrates of mechanically induced chemical reactions, weak intermolecular bonds (H-bonds, van der Waals two-body connections) should be the first to split under the mechanical action. This results in disordering or loosening of the corresponding surface layers.

At destruction of solids, a crack appears and develops. The peak of this moving crack is another point where intensification of chemical processes can arise. As a matter of course, significant shear stress in that point leads to high plastic deformation. Excess energy of this deformation excites vibration modes and produces the corresponding increase in temperature. Importantly, the fracture surface keeps the active centers for a long time after this fracture passing. For chemical reactions, favorable conditions are kept as well.

Deformation of solids is a response to the action of elastic energy impulses. At the impulse moment, different defects are formed (linear, point, planar, etc.) up to changes in structural type of a crystal lattice. The surface becomes amorphous or, in contrast, ordered and crystalline. Acceleration of diffusion occurs, which is also important for a chemical reaction. In the place of deformation, free radicals, coordination unsaturated atoms, and deformed interatomic bonds are formed as well as rearranged products.

Sliding contacts initiate the emission electrons, ions, and photons from the surface, which participate in sliding. The emission is often termed *microplasma* or *triboplasma*. The plasma action can result in molecular decomposition, as established for butane (Nakayama and Hashimoto 1996). Namely, the triboemission intensity of these three kinds of energetic particles depends on butane gas pressure. When the butane pressure is such that triboemission is maximal, the butane-originated polymeric products also form in the maximal amounts.

Chemical consequences of mechanical treatment are the tenor of this book. Besides chemical aspects, physical results are also important because they often define the chemical behavior of organic substances. First, the rate of dissolution of a solid compound increases with increasing surface area of the solid. Because bioavailability is related to dissolution kinetics and membrane permeability, the bioavailability of poorly soluble pharmaceutical or diagnostic compounds in many instances can also be increased via a reduction in the particle size. Importantly, in many instances it is especially desirable to have methods to reduce the size of pharmaceutical particles because many bulky pills containing small molecule drugs are poorly soluble in water or gastric fluids. Successful production of small particles can result in acceleration of the corresponding chemical reactions, and in respect to drugs, they can have faster therapeutic onset.

In the pharmaceutical and other industries, milling is a frequent method used for production of fine and ultrafine (nano) particles. The milling process typically involves charging grinding bodies to the milling chamber together with the material to be ground. In the case of wet milling, typically the material to be ground is added to the mill as a slurry comprised of a solid suspended in a liquid. Often, a surfactant is added to stabilize this slurry. Poly(ethylene glycol) stearate (a waxy material with a melting point of 33–37°C) is proposed as such a surfactant. It not only acts as a binder and increases the physical resistance of drug-ready forms, but also when these forms melt in the mouth, the surfactant provides rapid and complete solubilization of the drug (Abdelbary et al. 2004). For instance, ready-made acetaminophen tablets can be used with no need for swallowing. Many patients, seniors and children especially, find it difficult to swallow tablets and hard gelatin capsules. It is estimated that 50% of the population is affected by this problem. They do not take medications as prescribed. This results in a high incidence of noncompliance and ineffective therapy (Dobetti 2001).

Mechanochemical reactions are usually carried out in high-energy milling devices such as shaker mills, planetary mills, attritors, and vibration mills. All these apparatuses are considered in special engineering literature. The sources also point out such important factor as the ball-to-powder mass ratio (which is limited in ball mills of different types). As is well known, in a ball milling device an energy transfer to the milled powder occurs during repeated hits between balls and between balls and the vial walls. A new, near room temperature, high-energy ball milling technique has been developed (Cavalieri and Padella 2002; see also references therein). The method consists of enhancing mechanochemical effects promoted by the milling action through the insertion of a portion of liquid carbon dioxide in the milling vial. At each hit event, the energy is transferred from the milling device to the milled system. This promotes repeated microexplosive evaporation of liquid carbon dioxide, which is trapped between the ball and the vial wall. The described phenomena enhance the effectiveness of energy transfer and homogenization of a reacting mixture.

The size reduction depends on both the material properties and the mill performance. Regarding material properties, tensile modulus, hardness, and critical stress intensity factor should be determined first. These properties represent the resistance of the material to elastic deformation, plastic deformation, and crack propagation, respectively (Kwan et al. 2004).

For mill performance, several factors affect its efficiency, including the mill type, speed, or frequency; the shape, size, and material of the milling bodies; temperature; controlled atmosphere; and so on. Thus, the flat-end hardened steel vial (milling chamber) appears to be more efficient that the round-end vial made from the same material (Takacs and Sepelak 2004). The type of grinding items charged to the media mill is generally selected from any variety of dense, tough, hard materials, such as sand, stainless steel, zirconium silicate, zirconium oxide, yttrium oxide, glass, alumina, titanium, and the like. In situations involving either metal (oxide) contamination or shift in pH, polymeric grinding media are utilized.

Typically, the grinding items charged to the milling chamber have consisted of spherically shaped media milling beads. Spherically shaped grinding media have been thought the most mechanically stable form of hard grinding media because theoretically there are no edges to be rubbed or chipped off. When the agitating masses are too viscous or when there is a need to swiftly produce ultrafine particles, nonspherical items are used. They can be cubic, rectangular, hexagonal, rodlike, or ellipsoidal. Of course, these nonspherical items must have sufficient hardness and low friability to avoid chipping and crushing during grinding. For example, Daziel and others (2004) recommended use of items from polymeric resins, such as trademarked Derlin, Teflon, and some biodegradable polymers.

1.5 CONCLUSION

Chapter 1 provides an introduction to mechanochemical processes. The aim is to outline the main regularities governing transformations of mechanically activated organic compounds. Physical processes that affect these transformations are also discussed. The next chapters more specifically discuss the mechanically induced reactions of organic synthesis and the chemical transformations of organic participants of boundary lubrication. Good lubricity is important for the normal and prolonged work of devices and machines. Mechanically induced synthesis of the desired organic compounds is advantageous in the sense of yields and rates of formation. Also, it is often performed as solid-state reactions without organic solvents. This eliminates the need for solvent regeneration and makes the corresponding processes friendly from an environmental perspective. Summarily, this is preferred over expenses required for mechanical activation (say, for the electricity spent to rotate a mill).

REFERENCES

Abdelbary, G., Prinderre, P., Eouani, C., Joachim, J., Reynier, J.P., Piccerelle, Ph. (2004) *Int. J. Pharmaceutics* **278**, 423.

Beyer. M.K. (2003) *Angew. Chem. Int. Ed.* **42**, 4913.

Bisio, G., Cartesegna, M., Rubatto, G., Bistolfi, F. (2003) *Chem. Eng. Commun.* **190**, 177.

Bravi, M., Di Cave, S., Mazzarotta, B., Verdone, N. (2003) *Chem. Eng. J.* **94**, 223.

Cavalieri, F., Padella, F. (2002) *Waste Manage.* **22**, 913.

Dadali, A.A., Lang, J.M., Drickamer, H.G. (1994) *J. Photochem. Photobiol.* **84**, 203.

Dadali, A.A., Lastenko, I.P., Buchachenko, A.L. (1988) *Khim. Fiz.* **7**, 74.

Dadali, A.A., Lastenko, I.P., Danilov, V.G., Blank, V.D., Pisarenko, L.M. (1992) *Zh. Fiz. Khim.* **66**, 3076.

Daziel, S.M., Ford, W.N., Gommeren, H.J.C., Spahr, D.E. (2004) *Int. Pat. WO* 045585 A1.

Dobetti, L. (2001) *Pharm. Technol. Drug Delivery*, 44.

Drexler, R.K. *Nanosystems, Molecular Machinery Manufacturing and Computation* (Wiley, New York, 1992).

Gutmann, V., Resch, G. In: *Reactivity of Solids: Past, Present and Future*. Edited by Boldyrev, V.V. (Blackwell Science. Oxford, UK, 1996, p. 1).

Heinicke, G. *Tribochemistry* (Carl Hanser Verlag, Munich, Germany, 1984).

Hsu, S.M., Zhang, J., Yin, Zh. (2002) *Tribol. Lett.* **13**, 131.

Imamura, K., Senna, M. (1982) *J. Chem. Soc., Faraday Trans. 1* **78**, 1131.

Kajdas, Cz. (2001) *Tribol. Ser.* **39**, 233.

Kajdas, Cz. (2005) *Tribol. Int.* **38**, 337.

Kajdas, Cz., Furey, M.J., Ritter, A.L., Molina, G.J. (2002) *Lubr. Sci.* **14**, 223.

Kwan, Ch., Chen, Y.Q., Ding, Y.L., Papadopoulos, D.G., Bentham, A.C., Ghadiri, M. (2004) *Eur. J. Pharm. Sci.* **23**, 327.

Momose, Y., Iwashita, M. (2004) *Surface Interface Anal.* **36**, 1241.

Mulas, G., Loiselle, S., Schiffini, L., Cocco, G. (1997) *J. Solid State Chem.* **129**, 263.

Nakayama, K., Hashimoto, H. (1996) *Tribol. Int.* **29**, 385.

Nikulin, V.I., Pisarenko, L.M. (1985) *Izvest. Akad. Nauk, Ser. Khim.*, 151.

Ostwald, W. *Leherbuch der allgemeine Chemie, Bd.2 Stoechiometrie* (Engelmann, Leipzig, Germany, 1891).

Rapoport, N.Y., Zaikov, G.E. (1983) *Uspekhi Khim.* **52**, 1568.

Sakaguchi, M., Kashiwabara, H. (1992) *Colloid Polym. Sci.* **270**, 621.

Takacs, L. (2004) *J. Mater. Sci.* **39**, 4987.

Takacs, L., Sepelak, V. (2004) *J. Mater. Sci.* **39**, 5487.

Zarkhin, L.S., Burshtein, K.Ya. (1982) *Vysokomol. Soedin., Ser. B* **24B**, 695.

2 Mechanochromism of Organic Compounds

2.1 INTRODUCTION

Thermal action, solvation, and electric field effects can provoke color changes in organic materials. The names of these phenomena are *thermochromism*, *solvatochromism*, and *electrochromism*, respectively. A change in the chemical environment of organic chromophores such as pH is a classical case of chromotropism. Binding of specific biological targets by the chromophores is termed *affinochromism* or *biochromism* (Charych et al. 1996).

This chapter deals with mechanically induced color changes or *mechanochromism*. The aim of the chapter is to describe molecular transformations, inner crystal phenomena, and disordering or reorientation of monolayers provoked by mechanical effects. Mechanochromism is used to record and treat information and to study lubrication phenomena, the mechanically generated changes in molecular structures, or crystal packing (Todres 2004). This chapter considers simple molecules, crystals of organic metallocomplexes, and films formed by low- or high-weight organic compounds.

Thus, C_{60} fullerene films on mica change their low-temperature luminescent spectra on mechanical stress created by bending (Avdeenko et al. 2004). Monolayers created by Langmuir-Blodgett or self-assembly techniques (Ulman 1991) are particularly interesting. The ability to create organized ultrathin films using organic molecules provides systems with chemical, mechanical, and optical properties can be controlled for practical applications. In particular, polymerization of oriented mono- and multilayer films containing the diacetylene group has produced a variety of robust, highly oriented, and environmentally responsive films with unique chromatic properties (Bloor and Chance 1985). Mueller and Eckhardt (1978) reported an irreversible transition in a polydiacetylene single crystal induced by compressive stress, which resulted in coexisting blue and red phases. Nallicheri and Rubner (1991) described reversible mechanochromism of conjugated polydiacetylene chains embedded in a host elastomer that was subjected to tensile strain. Tomioka et al. (1989) induced reversible chromic transition by varying the lateral surface pressure of a polydiacetylene monolayer on the surface of water in a Langmuir–Blodgett trough, with the red form present at higher (compressive) surface pressures. These studies fixed the mechanochromism phenomena and opened ways to examine the molecular-level structural changes associated with the observed transitions.

In Chapter 2, different causes of organic mechanochromism are elucidated and considered, and corresponding representative examples are discussed. When relevant, technical applications of the phenomenon are highlighted.

2.2 MECHANICALLY INDUCED LUMINESCENCE

2.2.1 LUMINESCENCE CAUSED BY MECHANICAL GENERATION OF CRYSTAL DEFECTS OR CHANGES IN INTERMOLECULAR CONTACTS

Triboluminescence, the emission of light by solids when they are stressed or fractured, is a very common phenomenon. For a long time, it has been known that sugar shines if is triturated in the dark. According to literature estimates, 36% of inorganic, 19% of aliphatic, and 37% of aromatic compounds; 70% of alkaloids; and perhaps 50% of all crystalline materials are triboluminescent (Sweeting 2001 and references therein).

Although it remains an obscure phenomenon, the effect is generally explained (Sweeting et al. 1997) in terms of excitation of the molecule by an electric discharge between the surfaces of the fractured crystals. Indeed, emission of light, radio signals, electrons, and ions were clearly demonstrated at the moment of fracture under vacuum (Dickinson et al. 1984). Some materials, even molecular crystals or small molecules, do emit light without involvement of the surface, most likely by recombination of energetic defects during deformation or fracturing (Sweeting 2001). The discharge causes luminescence.

Triboluminescent compounds have received increasing attention partly because of a growing need for optical-pressure sensor devices and structural damage sensors (Sage and Bourhill 2001). These compounds are useful in the study of wear (Nakayama and Hashimoto 1995) and material fracture (Xu et al. 1999). Photochemistry resulting from triboluminescence is implicated in the mechanism of explosions (Field 1992).

A novel and potentially important application of the phenomenon has been proposed for the development of real-time damage sensors in composite materials (Sage et al. 1999). Specifically, the idea is based on the observation that light emission occurs when a composite material containing the triboluminescent molecule is damaged. Monitoring the wavelength and measuring the amount of emission yield information on the extent of the damage. During the development of such sensors, relationships between triboluminescence and solid-state photoluminescence were carefully studied, and some important features were revealed. For instance, certain pure organic crystals emit light on fracturing, but this light is significantly self-absorbed within the bulk of a damaged crystal (Duignan et al. 2002).

Early experiments indicated that the triboluminescent phenomenon was commonly observed in materials with noncentrosymmetric (noncentric) structures, which therefore were piezoelectric. For metal complexes with organic ligands, this feature is discussed in the literature as Zink's rule (Hocking et al. 1992; Zink 1978).

Several examples of triboluminescent molecules should be considered here with respect to their structural features. These examples differ in their centrosymmetry, that is, in order-disorder features of the molecular structures. Thus, *tris*(2-thenoyltrifluoro-acetone)europium(4,4'-dimethyl-2,2'-dipyridyl) [(tta)$_3$·Eu·dmdpy in Scheme 2.1] has a noncentrosymmetric crystal structure and disorder of the thyenyl rings and the trifluoromethyl groups (Zheng et al. 2002). Further, single crystals of a terbium complex (tta)$_3$·Eu·dmdpy   (fod)$_3$·Eu·phen

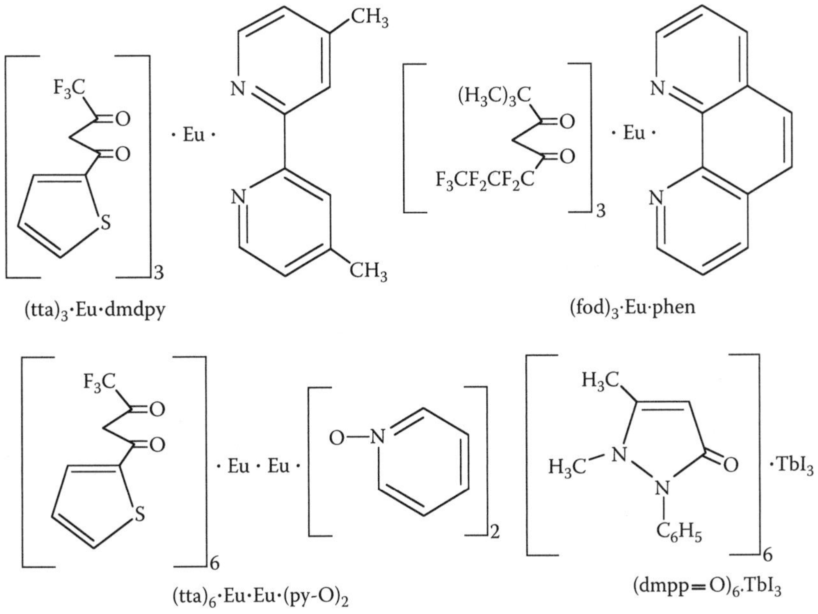

(tta)$_6$·Eu·Eu·(py-O)$_2$ (dmpp=O)$_6$·TbI$_3$

SCHEME 2.1

with five 2,6-dihydrobenzoic acid ligands coordinated to the metal center and containing two tetrabutylammonium cations exhibit green luminescence detectable by the naked eye when ground by hand in daylight. The spectrum exactly matches that observed by excitation using ultraviolet light in solution. This indicates that tribolu-minescence involves the same Tb^{3+} excited-state deactivation as that observed in photoluminescence. The terbium complex is a noncentrosymmetric ionic crystal (Soares-Santos et al. 2003). The authors noted that "given its triboluminescent behavior, the terbium compound is a potential candidate for development of optical sensors."

On the other hand, hexakis(2,3-dimethyl-1-phenylpyrazolone-5) terbium triiodide [(dmpp=O)$_6$TbI$_3$ in Scheme 2.1] is clearly centrosymmetric without any disorder (Clegg et al. 2002). Centrosymmetry has also been proven for a μ_2-(pyridine N-oxide) bridged binuclear europium(III) complex [(tta)$_6$·Eu·Eu·(py-O)$_2$ in Scheme 2.1] (Chen et al. 2002).

All the structures in Scheme 2.1 are brilliantly triboluminescent. Sparks are displayed when their crystals are cut or crushed in the dark and even with room illumination. For the noncentrosymmetrical europium mononuclear complex, the observed triboluminescent activity is ascribed to charge separation caused by piezo-electricity, which occurs when the nonsymmetric crystal is deformed or fractured. For the centrosymmetric terbium complex, the authors assumed that impurities create piezoelectric charge separation. For the other centrosymmetric species, (tta)$_6$·Eu·Eu·(py-O)$_2$ in Scheme 2.1, the authors experimentally excluded impurities by careful crystallization, and in their opinion, impurities were not responsible for the triboluminescence observed. The complex $(tta)_6 \cdot Eu \cdot Eu \cdot (py-O)_2$ is not ionic and could not gain the local piezoelectricity essential to its triboluminescent activity. However, there is disorder of all six thienyl rings and the trifluoromethyl groups. This disorder may provide a structural basis for charge separation by creating randomly distributed sites of slightly different ionization potentials and electron affinities at the faces of developing cracks. Some authors consider disorder as the essential condition for triboluminescent activity even in cases of centrosymmetric ionic crystals (e.g., see Xiong and You 2002). Such disorder may provide the local dissymmetry needed to support charge separation.

The mechanoluminescence of tris(6,6,7,7,8,8,8-heptafluoro-3,5-octanedione-2,2-dimethyl) europium phenanthroline complex [(fod)$_3 \cdot Eu \cdot$phen in Scheme 2.1] was also observed and studied with respect to the role of the two constituents, $-(fod)_3 \cdot Eu$ and phenathroline (phen) (Kazakov et al. 2003). When it is crushed separately, the phen crystal emits luminescence that is distinct from that of the (fod)$_3 \cdot Eu \cdot$phen complex, whereas the disintegration of (fod)$_3 \cdot Eu$ crystals is not accompanied by any luminescence. However, luminescence analogous to that from the (fod)$_3 \cdot Eu \cdot$phen complex does take place on grinding of phen mixture with (fod)$_3 \cdot Eu$. The inference is that the chromic effect observed is due to the sensitization of phen triboluminescence by the (fod)$_3 \cdot Eu$ part of the (fod)$_3 \cdot Eu \cdot$phen complex.

The complexes (piperidinium)[Eu(benzylacetyl methide)$_4$] and Tb(antipyrine)$_4$I$_3$ provide additional examples of exceptions to Zink's rule. Careful x-ray studies showed that they are centrosymmetric, but when they are ground, they display luminescence visible to the naked eye (Cotton and Huang 2003).

One special case of mechanochromism is provided by the complex *bis*[gold(1+) trifluoroacetate] containing uracilate (or methyluracilate) and *bis*(diphenylphosphino)methane as ligands, which forms helical crystals and in which there are weak gold–gold intermolecular contacts. When it is gently crushed under a spatula, the complex exhibits bright blue photoluminescence. The complex eliminates trifluoroacetic acid and becomes more linear, and this markedly enforces the gold–gold contacts (Lee and Eisenberg 2003).

Chakravarty and Phillipson (2004) compared triboluminescence from fractured sugar and terbium hexakis(antipyrine) triiodide. In both cases, fracture resulted in the breaking of bonds in such a manner that a sufficient number of bonding electrons (or negative ions) remained with the surface. At the same time, the opposite, positively charged, surface forms. After that, electrical discharges take place. Such electrical discharges are responsible for the detected light emission. At this moment, a potential difference of some hundreds of volts is attained.

One principal difference was noted between triboluminescence of sugar and the terbium complex. In the case of sugar, the pulses of light are much longer than the discharge event recorded. This is because the relatively long decay time of the excited forms of nitrogen that are created during an electrical discharge in air. In contrast, triboluminescence of the terbium complex commences prior to electrical discharge. It was demonstrated by Chakravarty and Phillipson (2004) that the terbium complex has high cathodoluminescence efficiency. This may explain the high intensity of triboluminescence of this material. The difference just discussed should be taken into account when highly triboluminescent materials are planned for use in practical applications.

2.2.2 LUMINESCENCE INDUCED BY SHOCK WAVES

Shock waves also generate triboluminescence in crystalline organic compounds such as *N*-isopropylcarbazole (Tsuboi et al. 2003). In this work, a novel technique was elaborated to generate a shock wave by means of laser irradiation onto a glass plate with a side (frontal to the laser beam) coated with a black pigment film. The film did not transmit light irradiation but propagated an intense pulsed-acoustic wave (i.e., shock wave) through it, which is sufficient to cause some fracture in crystalline *N*-isopropylcarbazole. The crystals are known as piezoelectric (Cherin and Burack 1966). Piezoelectrization of the fresh crystal surface was followed by an electrical discharge that produced luminescence.

Of course, shock waves can sometimes induce destruction of organic compounds, resulting in the formation of high-energy radicals that emit luminescence. The phenomenon was demonstrated for perfluoroalkanes, perfluoroalkyl amines, and perfluorotoluene (Voskoboinikov 2003).

2.2.3 LUMINESCENCE INDUCED BY PRESSURE

Some principal difference should be emphasized between pressure effects and the effects of fracture, shock wave, or shear. *Pressure* is the thrust distributed over a surface. From this definition, the difference is clear. The most important results of compression are reduction of molecular volumes of organochemicals, shortening the distances between molecules or molecular layers in crystals, structural phase transitions, and conformational changes.

Spectral changes of solids caused by pressure are termed *piezochromism*. There are various origins of piezochromism. The main causes are considered next based on representative examples.

2.2.3.1 Piezochromism as a Result of Structural Phase Transition

Luminescence of platinum (2,2′-bipyridyl) dichloride [Pt(bpy)Cl$_2$] depends on the surrounding pressure (Wenger et al. 2004). The red-colored [Pt(bpy)Cl$_2$] undergoes crystallographic phase transition at 1.7 GPa. This transition is associated with the conversion of the red complex to a denser yellow form.

The second illustrative example relates to coumarin 120 (7-amino-4-methyl-2H-1-benzopyran-2-one). This compound experiences structural phase transition and steady-state changes in its fluorescent behavior when the surrounding pressure is increased and then released quickly (Li et al. 2004). A new sample forms, but no chemical reaction takes place, and no new molecules are obtained. After compression to 7 GPa, the crystalline sample of coumarin 120 becomes amorphous and dense. This is because the molecules in the crystal become much closer to each other than they are at atmospheric pressure.

For organic crystalline compounds, it has been shown that high pressure disturbs weak interactions, which plays an important role in maintaining the crystal structure (Isaaks 1991; Zhu et al. 2001). During the process of decompression, the molecules move and relax to become crystallized again. Coumarin 120 bears the amino, carbonyl, and methyl groups and the oxygen atom within the nonaromatic ring. Because of the multipoints for hydrogen bonding and strong charge transfer inter-actions between molecules, some new crystalline structure is formed. This form is new crystallographically, but chemically it is a very weak (so-called van der Waals) complex comprised of the unchanged initial molecules. According to x-ray diffrac-tion analysis, the new structure is crystallographically not the same as that before squeezing (Li et al. 2004). On the contrary, during crystallization from melts or from solutions at atmospheric pressure, the process proceeds in an oriented way. Namely, the molecules move together from distances that can be considered infinitely far away. In this process, the molecules have enough time and distance to find perfect positions to form the crystal structure. Thus, only the same set of crystal structures may be formed at atmospheric pressure, and the fluorescence behavior of these crystals is the same as that of the starting coumarin 120.

2.2.3.2 Piezochromism as a Result of Intramolecular Charge Transfer

Two stable conformations of N-(1-pyrenylmethyl)-N-methyl-4-methoxyaniline, the linear and bent (V-like) forms are depicted in Scheme 2.2 (He et al. 2004).

Pressure provokes transition of the linear (extended) conformation into the bent (V-like) one. (The V-like form is more compact and occupies a smaller volume.) It is obvious that the V-like form is favorable with respect to intramolecular charge transfer from the donor (the aniline part) to the acceptor (the pyrene part). Such transfer is impossible in the case of the extended conformation caused by the large distance between the donor and acceptor moieties. The spectral changes observed reflect this stereochemical transition at elevated pressures.

In the spectral sense, there are three emitting sources: (1) local excited state of the pyrene moiety, (2) *inter*molecular charge transfer complex, and (3) *intra*molecular

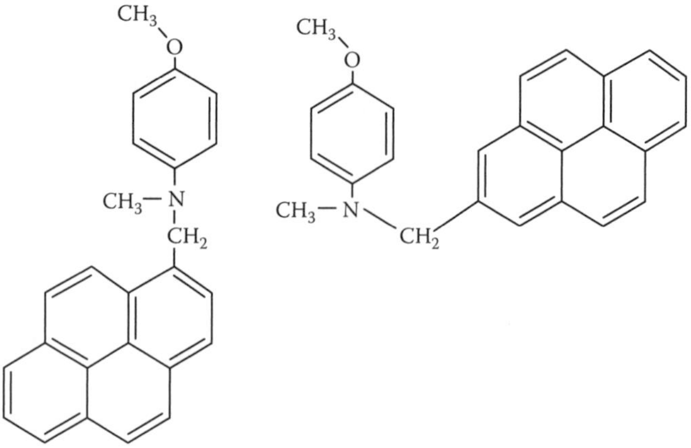

SCHEME 2.2

charge transfer complex. At normal pressure, the emission from the intermolecular complex is much larger than that from other sources. With the increase of pressure, the donor and acceptor moieties get much closer, and the emission of the intermolecular complex shifts greatly to the red side, and the relative efficiency drops significantly. Meanwhile, the emission from the intramolecular charge transfer complex increases with the surrounding pressure, and fluorescence efficiency becomes relatively higher. At higher pressure, the intermolecular and intramolecular complexes exist simultaneously, but their emission is observed at significantly different wavelength ranges. The intramolecular complex emission is enhanced by the increase of pressure (He et al. 2004).

2.2.3.3 Piezochromism as a Result of Changes in Interactions within a Single Crystal

The crystalline chloride salts of dioxorhenium(V) complexes containing ethylene diamine derivatives (Grey et al. 2001, 2002) or the crystalline tetrabutylammonium salts of the platinum(II) and palladium(II) complexes carrying thiocyanate or selenocyanate ligands (Grey et al. 2003) undergo blue shifts of band maxima and enhancement of luminescent intensities under pressure up to 0.4 MPa. The external pressure changes the interaction between ground and excited (emitting) states of the crystalline compounds, which results in the spectral effects observed.

There are examples that describe shortening of the distance between adjacent molecules in a crystal lattice, resulting in enhancement of intermolecular overlap of electron density. Oehzelt and coauthors (2003) studied the crystal properties of anthracene under high pressure. The unit cell volume and all the lattice parameters were decreased considerably at pressure up to 22 GPa. This pressure induces overlap of the electron densities between adjacent molecules in the crystal structure of anthracene. As a consequence, increased charge carrier mobility as well as red-shifted fluorescence and enhanced optical absorption become plausible. Both effects were observed (Hummer et al. 2003; Offen 1966). Yamamoto et al. (2003) observed the solid-state piezochromism of 5,8-*bis*(hexadecyloxy)-9,10-anthraquinone that was polymerized at the expense of its 1 and 4 positions. The corresponding film has orange-yellow color at ambient pressure but becomes dark red at about 1 GPa. In this case, piezochromism is reversible. At high pressure, shortening the face-to-face distance between the π-conjugated polymer planes increases the interlayer π–π interaction between the polymer molecules to cause piezochromism.

2.2.3.4 Changes in Luminescence of Organic Solutions as a Result of a Pressure-Induced Increase in Solvent Viscosity

Application of hydrostatic pressure does not modify the specific interaction between solute and solvent molecules but introduces geometrical restrictions for the solute. This can be exemplified with the case of luminescence observed at 10 and 40 MPa for a solution of *N*-phenyl-2,3-naphthalimide in triacetin (Hoa et al. 2004). The solvent triacetin is glycerol triacetate, and its viscosity increases strongly with pressure. Because the solvent becomes more viscous, the rotation of the *N*-phenyl ring is progressively hindered. Eventually, a pressure effect results in prevention of the formation of the excited state with planar geometry. Then, the excitation by light is localized on the naphthalimide moiety with no participation of the *N*-phenyl substituent. Under pressure, the fluorescence is therefore emitted at the short wavelength with a significantly enhanced intensity.

2.2.3.5 Periodic Changes in Fluorescence Intensity as a Result of Host–Guest Complexation/Decomplexation within Compression/Expansion Cycle

The case of periodic changes in fluorescence intensity as a result of host–guest complexation/decomplexation within the compression/expansion cycle is represented by the work of Ariga and coauthors (2005) (Scheme 2.3). In Scheme 2.3, the host is cyclophane connected to four cholic acid moieties through a flexible L-lysine spacer; the guest is potassium salt of N-(methylphenyl)naphthylamine sulfonic acid. On pressure, the host steroid moieties can be bent vertically; after that, the host grips the guest. As this takes place, the guest enhances its fluorescence intensity. Pressure release leads to unbending of the cholic moieties; the guest becomes free from the embraces, and its fluorescence intensity is diminished. Repeated compression and expansion induce periodic changes in the fluorescence intensity. This indicates a piezoluminescence effect through catch and release of the guest on the host dynamic cavity.

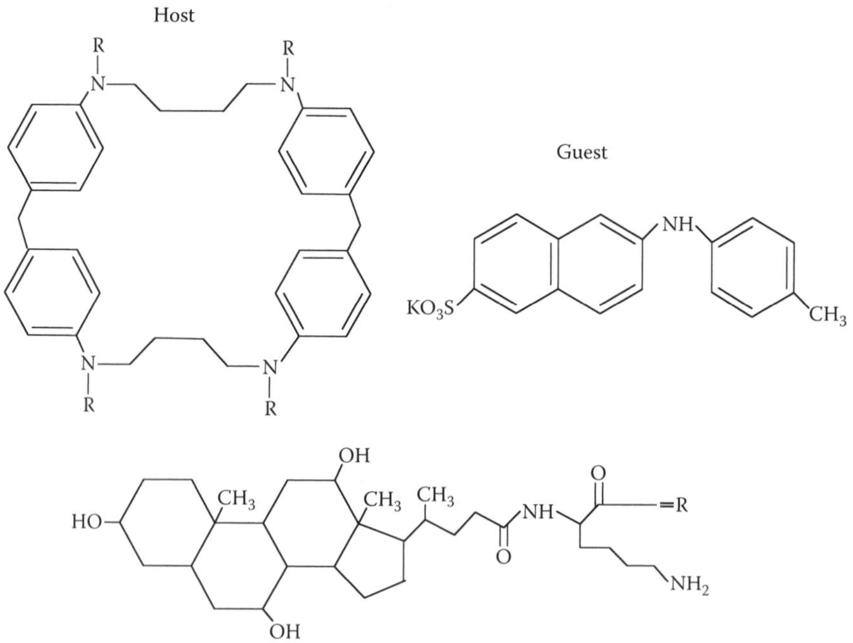

SCHEME 2.3

Experimentally, this case of piezoluminescence was observed at the air–water interface through dynamic molecular recognition driven by lateral pressure application. The experimental details are described in the work of Ariga and coworkers (2005). The authors assumed that the observed phenomenon can be used for controlled drug release and in designs of molecular sensing systems driven by mechanical stimuli.

2.3 COLORATION AS A RESULT OF RADICAL ION GENERATION ON MILLING

Mechanical processing (e.g., abrasion) of metallic surfaces causes the emission of electrons; this is known as the Kramer effect (Kramer 1950). The effect has been shown by the measurement of self-generated voltages between two metallic surfaces under boundary lubrication (Anderson et al. 1969; Adams and Foley 1975). Because the exoelectrons have a kinetic energy of about 1 to 4 eV (Kobzev 1962), they may initiate some chemical reactions. For instance, if the metal (with a surface that has been worked) is placed in an aqueous solution of acrylonitrile, the latter forms an abundant amount of an insoluble polymer, and the polymer amount accumulates after four days of contact. The same solution of acrylonitrile contains no polymer even after four months of contact with a piece of a nonworked metal. The following sequence of the reaction steps was proposed to explain the formation of polymer (Ferroni 1955):

$$H_2O + e \rightarrow H^{\cdot} + OH^-;\ H^{\cdot} + O_2 \rightarrow HO_2^{\cdot};\ HO_2^{\cdot} + H_2O \rightarrow H_2O_2 + \cdot OH;$$

$$CH_2 = CHCN + \cdot OH \rightarrow HOCH_2CH^{\cdot}CN;$$

$$HOCH_2CH^{\cdot}CN + CH_2 = CHCN \rightarrow HOCH_2CH(CN)CH_2CH^{\cdot}CN,\ etc.$$

When mechanical vibration of *bis*(pyridinium) salts (see Scheme 2.4) was conducted with a stainless steel ball in a stainless steel blender at room temperature

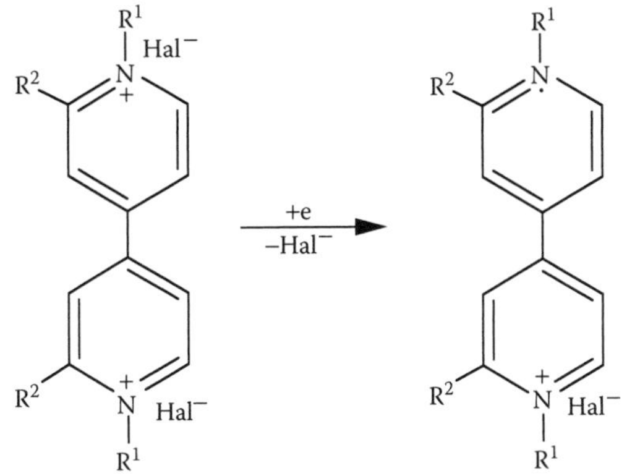

$$R^1 = Me,\ R^2 = CN,\ Hal = Br;\ R^1 = Ph,\ R^2 = H,\ Hal = Cl;$$
$$R^1 = Me,\ R^2 = H,\ Hal = Cl$$

SCHEME 2.4

under strictly anaerobic conditions, the powdery white surface of the dicationic salts turned deep blue-purple (Kuzuya et al. 1993). Nearly isotropic, broad, single-line electron spin resonance (ESR) spectra were recorded in the resulting powder. No ESR spectra were observed in any of the dipyridinium salts when mechanical vibration was conducted with a Teflon-made ball in a Teflon-made blender under otherwise identical conditions. When observed, the ESR signals were quickly quenched on exposure to air, and the starting dicationic salts were recovered.

Each of the resulting powders was dissolved in air-free acetonitrile, and the ESR spectra of the solutions were recorded after the material had been milled under anaerobic conditions. Analysis of hyperfine structure confirmed the formation of the corresponding radical cations depicted in Scheme 2.4.

A remaining electron in the molecular orbital of the radical cation is unpaired. With an unpaired electron, polarizability of a molecule increases, and its excitation by light is facilitated. This enhances the intensity of light absorption and shifts it to the region of higher wavelengths. The electrochromism is thus caused by the emission of exoelectrons from a metal-abraded surface.

2.4 BOND-BREAKING MECHANOCHROMISM

Mechanically induced bond breaking and skeletal isomerization can result in formation of colored forms of organic molecules. Several examples illustrate the origins of the phenomenon.

Spiropyrans contain a weak bond between the nodal carbon atom and the ethereal oxygen, and this bond is easily disrupted on photoirradiation, selective (polar) solvation, or thermal action. The isomerization is accompanied by a change of the initial yellow color to blue (Scheme 2.5).

Scheme 2.5 was drawn according to the principle of maximal structural conservation. Reversible transformation from the s-*cis* butadiene fragment (as depicted on the right side of Scheme 2.5) to the s-*trans* one is possible but not included for the sake of clarity.

The same color changes are also observed on grinding of the spiropyran depicted on Scheme 2.5 (Tipikin 2001), and the mechanically induced coloration is reversible. Oxidation or other chemical reactions in air do not cause the coloration. The sample warms by not more than 6 K during grinding, and as the melting point of the compound is 170°C, this temperature increase cannot lead to global melting of the sample and isomerization. Moreover, at liquid nitrogen temperature, the mechanochromic

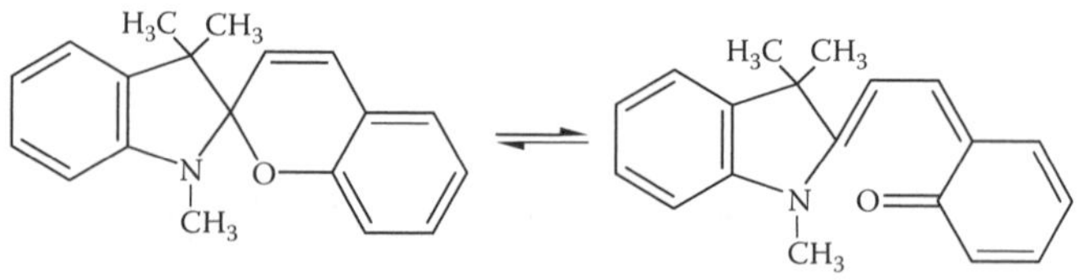

SCHEME 2.5

effect is enhanced. According to quantum mechanical calculation (Aktah and Frank 2002), almost simultaneous intramolecular electron transfer and bond disruption can occur on mechanical stress. In this case, mechanochromism is conditioned by formation of defects in the rigid matrix. The spiropyran molecules are excited in the surroundings of the defects and undergo ring opening to give the quinoid (colored) molecule, which is trapped in the crystal by lattice forces.

It would be interesting to check the possibility of generating the colored form on grinding, not of the pure spiropyran depicted on Scheme 2.5, but of the spiropyran included in a cyclodextrin cavity. Such inclusion has been shown to prevent the ring-opening isomerization of the spiropyran under thermal conditions (Sueishi and Itami 2003).

In other cases, bond breaking gives rise to radicals that cause oscillating coloration. Scheme 2.6 represents mechanochemically induced formation of two radicals from 2,2'-*bis*[4-(dimethylamino)phenyl]-1,3-indandiones (Pisarenko et al. 1987, 1990).

In the radicals from Scheme 2.6, an unpaired electron is distributed between the two carbonyls and aromatic rings, resulting in coloration. In the solid phase, the mechanochemical reaction is reversible. Onium substituents in the condensed benzene rings (trimethylammonium or 2,4,6-triphenylpyridinium) increase the thermodynamic stability of the radicals, shift the equilibrium to the colored products, and intensify the mechanochromic effect.

Studies of dibenzofuranones revealed structural effects on mechanochromism. The skeleton difference between the β-β′ and α-α′ isomers is depicted on Scheme 2.7 (Mori et al. 1995).

The β-β′ isomer from Scheme 2.7 was reported to exhibit piezochromism when the solid was rubbed or pressed (Loewenbein and Schmidt 1927). Reversible homolytic rupture of the β-β′ bond gives blue-colored radicals, which recombine to give the colorless dimer (Ohkada et al. 1992).

On the other hand, the α-α′ isomer from Scheme 2.7 does not undergo homolysis under the same conditions (Mori et al. 1995). The crystal structure of the β-β′ isomer and conformational analysis of both isomers, show that the β-β′ isomer experiences strong internal strain and restriction of rotation around the exocyclic bond, whereas

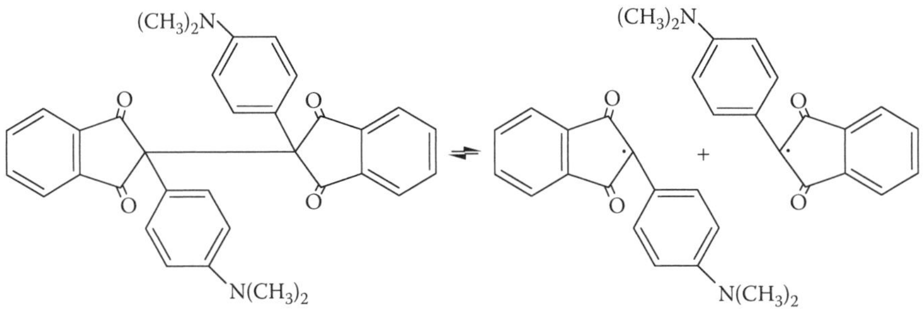

SCHEME 2.6

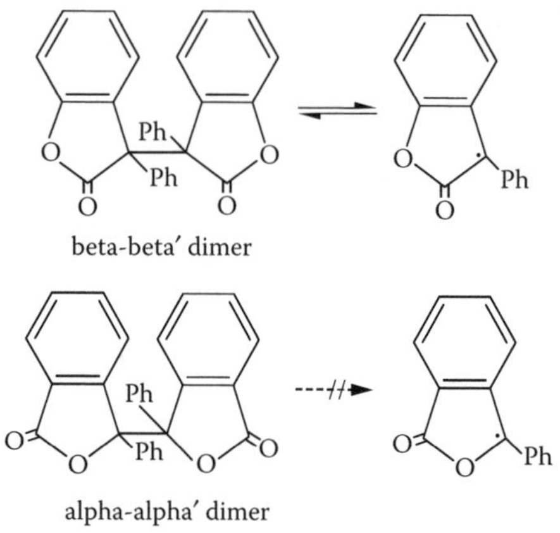

beta-beta′ dimer

alpha-alpha′ dimer

SCHEME 2.7

such restriction is absent in the α-α′ isomer. The former is less stable and much more sensitive to rubbing and pressing, resulting in mechanochromism.

One interesting (but not completely explained) case of coloring resulting from the bond scission was observed on grinding polymeric oxovanadium(IV) complexes carrying Schiff base ligands (Kojima et al. 1995; Nakajima et al. 1996; Tsuchimoto et al. 2000). For instance, the orange vanadyl complex with N,N′-disalicylidene-(R,R′)-1,2-diphenyl-1,2-ethanediamine turns green during grinding. The compound has a linear polymeric structure containing an infinite chain of ···V=O···V=O··· bonds, and these are cleaved to yield monomer species. The green product obtained by grinding reverts to orange when moistened with small amounts of acetonitrile or acetone or by exposure to the acetonitrile vapor, which suggests a reformation of the thermodynamically stable polymeric structure. This color change can be repeated. It is clear that the mechanochromic rearrangement starts at the lattice defect sites; formation of solvatocomplexes may also play an important role in crystal packing. However, the full understanding of the driving force of such coloring is still waiting for additional studies. The cause of the orange-to-green change might be connected with coordinative unsaturation in the monomeric complex formed after grinding.

Grinding or milling of poly(methylmethacrylate) in a steel apparatus causes mechanical destruction of the polymer, which is accompanied by luminescence (see the review part of the 2002 article by Zarkhin). However, the processes of milling and steel friction are not equivalent to real mechanical destruction. The points are exoelectron emission. The elastic energy of the broken chain transforms into electron excitation, which is centered on the carbonyl chromophore of the polymer fragments. A short-term emission induced at the instant of mechanical fracture was detected. Kinetically, mechanoluminescence develops through two phases: The first stage is related to the propagation of the main crack at a subsonic velocity, whereas the second stage is related to the cracking of a freshly formed fracture surface on its rapid cooling and a concomitant glass transition. Processes of mechanoluminescence emission and crack initiation/propagation are synchronous. The rupture of polymer chains exclusively causes this behavior. The phenomenon can be used for measuring the initiation and growth of cracks in polymers.

Assuming that pressure is a kind of mechanical action, the pressure-induced *cis–trans* isomerization of poly[(*p*-methylthiophenyl)acetylene] should be considered (Huang et al. 2004). The compression (up to 2 MPa) "partially" breaks the *cis* C=C double bond to create two unpaired electrons with the formation of a (—C̣—C̣—) biradical ordinary bond. Pairing of the two electrons leads to (—C=C—) double bond again but in the different, *trans*, configuration. The *trans* form is more stable because of π-conjugation through the whole bone chain of the polymer. Accordingly, a new absorption band is observed at a fairly longer wavelength, shifted to the visible region. The length of the *trans* π-conjugated sequences generated by the compression of the polymer was estimated as about 40 carbon–carbon links. The authors ascribed the driving force of the pressure-induced *cis–trans* isomerization to the difference in the effective molecular volumes between the *cis* and *trans* isomers. Namely, the molecular volume of the *trans* isomer is about one half compared with that of the *cis* isomer. Some part of (—C̣—C̣—) bonds in the polymer remains in the biradical state: The ESR spectrum shows coexistence of *cis* and *trans* biradicals. "It seems that the methylthio-substituted polyacetylene with high *cis* content may function as the origin of the radical spin necessary for forming an organo-magnetic material" (Huang et al. 2004).

2.5 SPECTRAL CHANGES AS A RESULT OF MECHANICALLY INDUCED REORGANIZATION OF CRYSTAL PACKING

One novel thioindigoid, 11-(3′-oxodihydrobenzothiophen-2′-ylidene)cyclopenta-[1,2-b:4,3-b′]dibenzothiophene, was found to undergo a color change from red to black when the powders were ground in a mortar with a pestle (Mizuguchi et al. 2003). Interestingly, the color is recovered when the powder (after grinding) is heated at about 280°C for 2 h or immersed in organic solvents for several minutes. Based on data from x-ray diffraction, electron spectroscopy, and molecular orbital calculations, the authors pictured the following sequence of events that led to the color change: Mechanical stress initiates partial slipping of the thioindigoid molecules along the stacking axis in the crystal. This shortens an interplanar distance along the molecular stack. An additional (new) wide band appears around 750 nm to make the color black. The new band has been interpreted as arising from excitonic interactions between transition dipoles based on the reorganized molecular arrangement during mechanical shearing (Mizuguchi et al. 2003). The recovery of the color from black to red is explained by the disturbed lattice, corresponding to a metastable state, relaxing and returning to a stable state. On heating, the metastable-to-stable phase transition happens because of the lattice vibration. Immersion in organic solvents loosens the crystal lattice, thus allowing the molecules to slide or rotate to find an energetically more stable site in the initial state. This picture describes a new route to mechanochromic transition at the expense of reorganization in crystal packing.

2.6 SPECTRAL CHANGES AS A RESULT OF MECHANICALLY INDUCED STRUCTURAL PHASE TRANSITION

Changes in optical thickness of self-assembled polymeric photonic crystals are manifestations of mechanochromism resulting from structural phase transition. Mechanochromic films, which respond to deformation by color alteration, are examples.

Thus, co-extruded AB elastomer multilayers reflect light from the visible to infrared regions. In a review, Edrington and coauthors. (2001) cited spectrochemical results of biaxial compression on multilayers fabricated by casting polystyrene and polyvinyl alcohol. At an applied pressure up to 2 MPa, a shift of 65 nm in the peak reflectivity was found. Mao et al. (1998) described applications of such self-assembled block polymeric materials to optical switches.

Caprick and others (2000a, 2000b, 2004), Burns and coauthors (2001), and Seddon et al. (2002) studied the shear-forced nanoscale mechanochromism of poly-dicetylene monolayers on an atomically flat silicon oxide support. The mechano-chromism was observed as irreversible transformation of the initial long-wave absorbed form into the short-wave absorbed conformation, corresponding to a change from the blue to red. This blue-to-red transition is dependent on the shear forces exerted on the pendant side chains. The transformation is also facilitated by defects in the support lattice. Structurally, the side chains are pushed toward the surface according to Scheme 2.8

The initial (blue) form contains the polymer backbone in the planar all-*trans* geometry, in which the side chains are in the same plane as the backbone. This

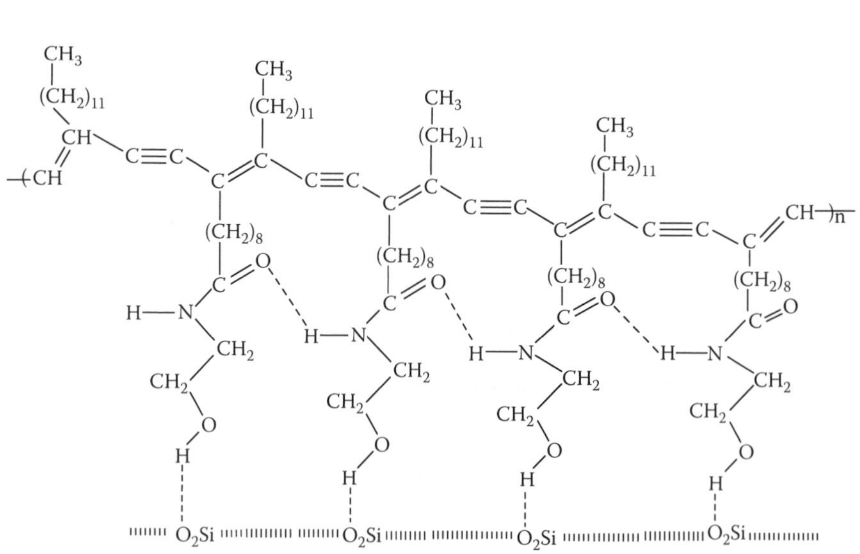

SCHEME 2.8

geometry permits extended, continuous conjugation between the double and triple bonds of the backbone that runs in parallel with the support surface. Shear action leads to rotation around the ordinary carbon–carbon bonds of the polymer backbone, thus changing the backbone planarity. The out-of-plane conformation of the side chains is achieved, and the conjugation in the backbone is disrupted. This shortens the conjugation length and evokes the hypsochromic shift in the absorption spectrum of the film. The same phenomenon was observed when mica (Caprick et al. 1999) and gold (Mowery et al. 1999) were used as the supports for Langmuir-deposited monolayer films.

The packing of the alkyl side chain and hydrogen bonding of the head groups jointly restrict the torsion mobility of the polymer backbone. The irreversibility of the transition observed indicates the greater stability of the red phase compared to the blue one.

Importantly, the friction in the red (bent) transformed regions increases up to 100% with respect to the blue regions and appears to correlate directly with the compression or reorientation of the side chains toward the surface. This mechanochromism can be used for elucidation of disordering, film defects, or shear forces operating in coated parts of technical devices.

Let us consider now mechanochromism of liquid crystalline linear polyacetylenes. One structure of this type is $H-(CH_2)_m-C{\equiv}C-C{\equiv}C-(CH_2)_8-[p-C(O)O-C_6H_4-C_6H_4OC(O)-p']-(CH_2)_8-C{\equiv}C-C{\equiv}C-(CH_2)_m-H$. These liquid crystals can serve as strain and pressure sensors (Angkaew et al. 1999). Mechanical action changes the liquid crystal orientation. Second-harmonic optical generation takes place. This kind of mechanochromism is observed if an electric field is applied to the mechanically stressed samples. Liquid crystals are oriented in the external electric field, and this ordering is disturbed when the field is eliminated. The long-axial molecules relax toward their equilibrium orientational order. The decay of the mechanically induced second-harmonic signal in time after the electric field has been switched off is a direct measure of the in-plane order rupture (Jerome et al. 2002).

In the absence of electric field, stretching or rubbing leads to the blue-to-red change of the polyacetylene films placed on top of quartz slide, and optical micrographs of the polymer coated on glass fibers show different morphology for the blue and red phases. The mechanochromic changes correspond to the structural phase transition from the original blue liquid crystalline phase to the red liquid crystalline phase. This mechanochromic transition is partially irreversible because of residual strain, and the chemical structural factors, such as alkyl spacer length, play an important role in controlling the optical properties.

As established, the organization in the liquid crystalline matrix is sensitive to the nature of chiral dopants present (Solladie and Zimmermann 1984). The pitch of a doped cholesteric phase can be changed by configurational inversion of the chiral dopant. Scheme 2.9 illustrates one pair as an example, namely, 4-pentyloxy-4'-biphenylcarbonitrile (M15 liquid crystal) with octahydrodimethyl biphenathrylidene as a dopant (van Delden et al. 2002).

The authors used the light beam directed at a 45° angle toward the film of the liquid crystal with the dopant. Reflection depends on the pitch of the liquid crystalline

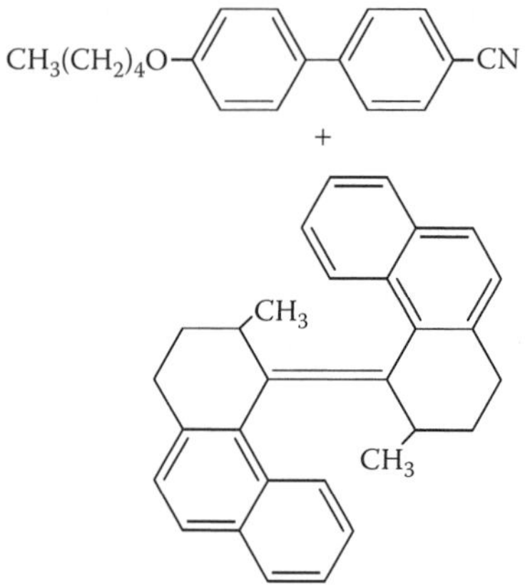

SCHEME 2.9

matrix. When this doped system is constantly fueled with photon energy from ultra-violet irradiation under appropriate thermal conditions, the diphenanthrylidene begins to rotate around the exocyclic double bond. This rotation entails changes in arrangement of the cholesteric liquid crystalline film, resulting in larger pitch and a red shift of the reflection wavelength. By tuning of the diphenanthrylidene rotary motion, all colors of the liquid crystalline film can be generated. The phenomenon just described seems to be promising for technical applications; however, further studies are needed. This was pointed out by the authors (van Delden et al. 2002, 2003).

Phase transformation, crystalline to amorphous, occurs when the free energy of the crystal is raised above that of the amorphous phase (Crowley and Zografi 2002). Grinding can increase the crystal free energy. At the same time, grinding can disrupt networks formed at the expense of hydrogen bonding of weak coordination. Scheme 2.10 puts two such examples forward. In both cases, grinding of crystalline solids provides the monomers in an amorphous state. The initial color is changed. The ground products have a high propensity to revert to the corresponding crystalline form, thereby reducing the free energy of the system. As this takes place, the initial color is restored. According to the authors, grinding may fix the tautomeric form of the monomer (Sheth et al. 2005; Jeragh and El-Dissouky 2004); see Scheme 2.10.

In conclusion, possible causes of amorphization deserve to be considered. Sheth et al. (2005) conducted their experiments with the top reaction of Scheme 2.10 at −195.8°C (in a bath of boiling liquid nitrogen). The sample under investigation was found not to undergo chemical degradation on grinding. Its melting point is about 200°C. Consequently, melting cannot be the cause of amorphization. At the same time, supercritical pressure really can give rise to Born instability and lattice collapse. In addition, as noted, mechanical action indeed raises the total energy of the sample and generates a driving force for the transition of crystals into the amorphous phase.

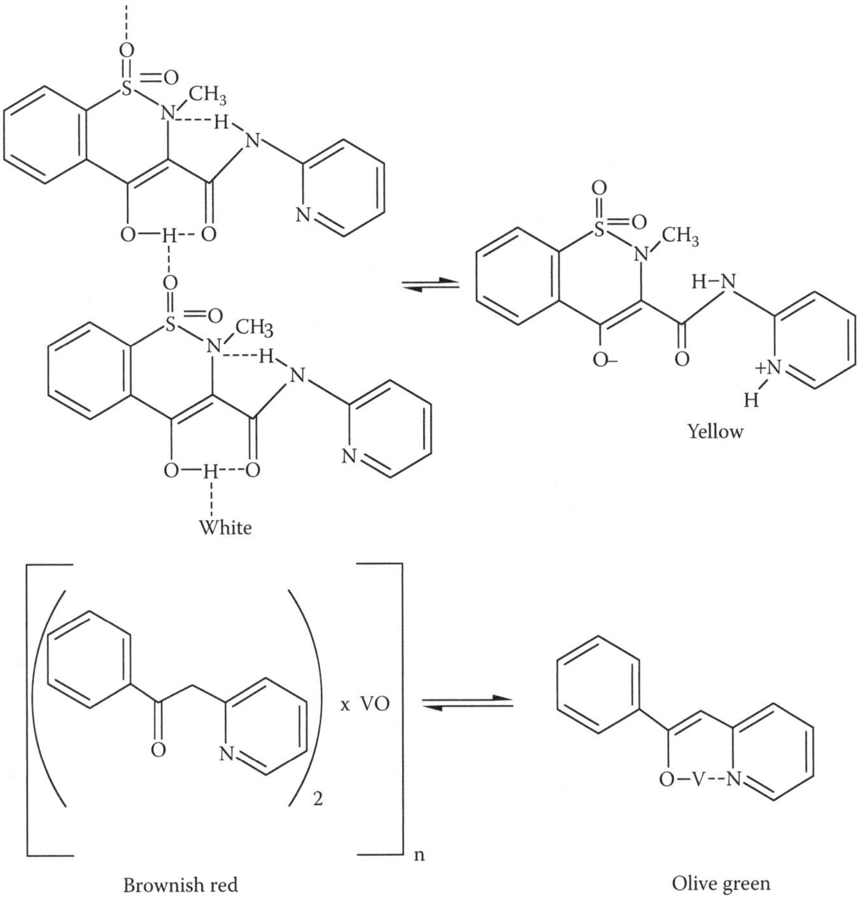

SCHEME 2.10

2.7 CONCLUSION

As seen from the material of Chapter 2, mechanochromism is a growing part of organic chemistry and the chemistry of materials. Many questions will arise during the development of information on mechanochromism. For example, it would be interesting to study the dependence of mechanochromism on temperature. That is important for the design of damage sensors working in various climatic conditions. Solid organic compounds such as α-lactose monohydrate (Hassanpour et al. 2004) show a significant decrease in the extent of breakage when the surrounding temperature is decreased. In other words, the mechanical energy utilization can depend on the temperature. Specific surface area per unit expended energy decreases as temperature decreases.

Our consideration has been made on the molecular level, and this chapter describes the conversion of mechanical energy into chemical driving force. Knowledge of the molecular transformations that cause mechanochromism provides access to new molecular systems that can be interesting both academically and technologically.

REFERENCES

Adams, A.A., Foley, R.T. (1975) *Corrosion* **31**, 84.

Aktah, D., Frank, I. (2002) *J. Am. Chem. Soc.* **124**, 3402.

Anderson, T.N., Anderson, J.L., Eyring, H. (1969) *J. Phys. Chem.* **73**, 3652.

Angkaew, S., Cham, P.M., Lando, J.B., Thornton, B.P., Brian, P., Ballarini, R., Mullen, R.L. (1999) *57th Annual Technical Conference — Society of Plastic Engineers* **2**, 1760.

Ariga, K., Nakanishi, T., Terasaka, Yu., Tsuji, H., Sakai, D., Kikuchi, J-I. (2005) *Langmuir* **21**, 976.

Avdeenko, A., Gorobchenko, V., Zinoviev, P., Silaeva, N., Zoryanskii, V., Gorbenko, N., Pugachev, A., Churakova, N. (2004) *Fiz. Nizkikh Temp.* **30**, 312.

Bloor, D., Chance, R.R. *Polyacetylenes: Synthesis, Structure, and Electronic Properties* (Martinus Nijhoff, Dordrecht, The Netherlands, 1985).

Burns, A.R., Carpick, R.W., Sasaki, D.Y., Shelnutt, J.A., Haddad, R. (2001) *Tribol. Lett.* **10**, 89.

Carpick, R.W., Sasaki, D.Y., Burns, A.R. (1999) *Tribol. Lett.* **7**, 79.

Carpick, R.W., Sasaki, D.Y., Burns, A.R. (2000a) *Langmuir* **16**, 1270.

Carpick, R.W., Sasaki, D.Y., Burns, A.R. (2000b) *Polym. Preprints (Am. Chem. Soc., Div. Polym. Chem.)* **41**, 1458.

Carpick, R.W., Sasaki, D.Y., Marcus, M.S., Eriksson, M.A., Burns, A.R. (2004) *J. Phys.: Condens. Matter* **16**, R679.

Chakravarty, A., Phillipson, T.E. (2004) *J. Phys. D: Appl. Phys.* **37**, 2175.

Charych, D., Cheng, Q., Reichert, A., Kuziemko, G., Stroh, M., Nagy, J., Spevak, W., Stevens, R.C. (1996) *Chem. Biol.* **3**, 113.

Chen, X.-F., Zhu, X.-H., Xu, Ya.-H., Raj, S.Sh.S., Fun, H.-K., Wu, J., You, X.-Z. (2002) *J. Coord. Chem.* **55**, 421.

Cherin, P., Burack, M. (1966) *J. Phys. Chem.* **70**, 1470.

Clegg, W., Bourhill, G., Sage, I. (2002) *Acta Crystallogr., Sect. E: Structure Reports Online* **E58**, m159.

Cotton, F.A., Huang, P. (2003) *Inorg. Chem. Acta* **346**, 223.

Crowley, K.J., Zografi, G. (2002) *J. Pharm. Sci.* **91**, 492.

Dickinson, J.T., Jensen, L.C., Jahan-Latibari, A. (1984) *Vacuum Sci. Technol. A* **2A**, 1112.

Duignan, J.P., Oswald, I.D.H., Sage, I.C., Sweeting, L.M., Tanaka, K., Ishihara, T., Hirao, K., Bourhill, G. (2002) *J. Lumin.* **97**, 115.

Edrington, A.C., Urbas, A.M., DeRege, P., Chen, C.X., Swager, T.M., Hadjichristidis, N., Xenidou, M., Fetters, L.J., Joannopoulos, J.D., Fink, Y., Thomas, E.L. (2001) *Adv. Mater. (Weinheim, Ger.)* **13**, 421.

Ferroni, E. (1955) *Res. Corres., Suppl. Res. (London)*, **8**, S49.

Field, J.E. (1992) *Acc. Chem. Res.* **25**, 489.

Grey, J.K., Butler, I.S., Reber, Ch. (2002) *J. Am. Chem. Soc.* **124**, 11699.

Grey, J.K., Butler, I.S., Reber, Ch. (2003) *Inorg. Chem.* **42**, 6503.

Grey, J.K., Triest, M., Butler, I.S., Reber, Ch. (2001) *J. Phys. Chem. A* **105**, 6269.

Hassanpour, A., Ghadiri, M., Bentham, A.C., Papadopoulos, D.G. (2004) *Powder Technol.* **141**, 239.

He, L., Xiong, F., Li, Sh., Gan, Q., Zhang, G., Li, Y., Zhang, B., Chen, B., Yang, G. (2004) *J. Phys. Chem. B* **108**, 7092.

Hoa, G.H.B., Kossanyi, J., Demeter, A., Biczok, L., Berces, T. (2004) *Photochem. Photobiol. Sci.* **3**, 473.

Hocking, M.B., VandervoortMaarschalk, F.W., McKiernan, J., Zink, J.I. (1992) *J. Lumin.* **51**, 323.

Huang, K., Mawatari, Ya., Tabata, M., Sone, T., Miyasaka, A., Sadahiro, Yo. (2004) *Macromol. Chem. Phys.* **205**, 762.

Hummer, K., Pushnig, P., Ambrosch-Draxl, C., Oezelt, U., Heimel, G., Resel, R. (2003) *Synth. Met.,* **137**, 935.

Isaaks, N.S. (1991) *Tetrahedron* **47**, 8463.

Jeragh, B.J.A., El-Dissouky, A. (2004) *Transition Met. Chem.* **29**, 579.

Jerome, B., Schuddeboom, P.C., Meister, R. (2002) *Europhys. Lett.* **57**, 389.

Kazakov, V.P., Ostakhov, S.S., Rubtsova, O.V., Antipin, V.A. (2003) *Khim. Vysokikh Energii* **37**, 156.

Kobzev, N.I. *Exoelectronic Emission* (Foreign Publishing House, Moscow, Russia, 1962).

Kojima, M., Nakajima, K., Tsuchimoto, M., Treichel, P.M., Kashino, S., Yoshikawa, Yu. (1995) *Proc. Jpn. Acad., Ser. B* **71B**, 175.

Kramer, J. *Der Metallische Zustand* (Vanderhock and Ruprecht, Goettingen, Germany. 1950).

Kuzuya, M., Kondo, Sh.-I., Murase, K. (1993) *J. Phys. Chem.* **97**, 7800.

Lee, Y.-A., Eisenberg, R. (2003) *J. Am. Chem. Soc.* **125**, 7778.

Li, H., He, L., Zhong, B., Li, Y. , Wu, Sh., Liu, J., Yang, G. (2004) *Chem. Phys. Chem.* **5**, 124.

Loewbein, A., Schmidt, H. (1927) *Chem. Ber.* **60**, 1851.

Mao, G., Wang, J., Ober, C., Brehmer, M., O'Rourke, M., Thomas, E.L. (1998) *Chem. Mater.* **10**, 1538.

Mizuguchi, J., Tamifuji, N., Kobayashi, K. (2003) *J. Chem. Phys. B* **107**, 12635.

Mori, Yu., Niwa, A., Maeda, K. (1995) *Acta Crystallogr., Sect. B: Struct. Sci.* **B51**, 61.

Mowery, M.D., Kopta, S., Ogletree, D.F., Salmeron, M., Evans, C.E. (1999) *Langmuir* **15**, 5118.

Mueller, H., Eckhardt, C.J. (1978) *Mol. Cryst. Liq. Cryst.* **45**, 313.

Nakayama, K., Hashimoto, H. (1995) *Wear* **185**, 183.

Nakajima, K., Kojima, M., Azuma, Sh., Kazahara, R., Tsuchimoto, M., Kubozono, Yo., Maeda, H., Kashino, S., Ohba, Sh., Yoshikawa, Yu., Fujita, J.U. (1996) *Bull. Chem. Soc. Jpn.* **69**, 3207.

Nallicheri, R.A., Rubner, M.F. (1991) *Macromolecules* **24**, 517.

Oehzelt, M., Heimel, G., Resel, R., Pusching, P., Hummer, K., Ambrosh-Draxl, C., Takemura, K., Nakayama, A. (2003) *Material Research Society Symposium Proceedings, 771 (Organic and Polymeric Materials and Devices)*, 219.

Offen, H.W. (1966) *J. Chem. Phys.* **44**, 699.

Ohkada, J., Mori, Yu., Maeda, K., Osawa, E. (1992) *J. Chem. Soc., Perkin Trans. 2*, 59.

Pisarenko, L.M., Gagarina, A.B., Roginskii, V.A. (1987) *Izvest. Akad. Nauk SSSR, Ser. Khim.*, 2861.

Pisarenko, L.M., Nikulin, V.I., Blagorazumnov, M.P., Neilands, O., Paulins, L.L. (1990) *Izvest. Akad. Nauk SSSR, Ser. Khim.*, 1525.

Sage, I., Badcock, R., Humberstone, L., Geddes, N., Kemp, M., Bourhil, G. (1999) *Smart Mater. Struct.* **8**, 504.

Sage, I.C., Bourhill, G. (2001) *J. Mater. Chem.* **11**, 231.

Seddon, A.M., Patel, H.M., Burkett, S.L., Mann, S. (2002) *Angew. Chem., Int. Ed.* **41**, 2988.

Sheth, A.R., Lubach, J.W., Munson, E.J., Muller, F.X., Grant, D.J.W. (2005) *J. Am. Chem. Soc.* **127**, 6641.

Soares-Santos, P.C.R., Noguera, H.I.S., Paz, F.A.A., Sa Ferreira, R.A., Carlos, L.D., Klinowski, Ja., Trinidade, T. (2003) *Eur. J. Inorg. Chem.*, 3609.

Solladie, G., Zimmermann, R.G. (1984) *Angew. Chem., Int. Ed.* **23**, 348.

Sueishi, Y., Itami, Sh. (2003) *Z. Phys. Chem.* **217**, 677.

Sweeting, L. (2001) *Chem. Mater.* **13**, 854.

Sweeting, L.M., Reingold, A.L., Gingerich, J.M., Rutter, A.W., Spence, R.A., Cox, C.D., Kim, T.J. (1997) *Chem. Mater.* **9**, 1103.

Tipikin, D.S. (2001) *Zh. Fiz. Khim.* **75**, 1876.

Todres, Z.V. (2004) *J. Chem. Res.*, 89.

Tomioka, Y., Tanaka, N., Imazeki, S. (1989) *J. Chem. Phys.* **91**, 5694.

Tsuboi, Ya., Seto, T., Kitamura, N. (2003) *J. Phys. Chem. B* **107**, 7547.

Tsuchimoto, M., Hoshina, G., Yoshioka, N., Inoue, H., Nakajima, K., Kamishima, M., Kojima, M., Ohba, Sh. (2000) *J. Solid State* **153**, 9.

Ulman, A. *Introduction to Organic Films from Lagmuir-Blodgett to Self-Assembly* (Academic Press, New York, 1991).

van Delden, R.A., Koumura, N., Harada, N., Feringa, B.L. (2002) *Proc. Natl. Acad. Sci. USA* **99**, 4945.

van Delden, R.A., ter Wiels, M.K.J. Koumura, N., Feringa, B.L. In: *Molecular Motors*, Edited by Schliwa, M. (Wiley-VCH, Weinheim, Germany, 2003, p. 559).

Voskoboinikov, I.M. (2003) *Fizika Goreniya Vzryva* **39**, 105.

Wenger, O.S., Garcia-Revilla, S., Gudel, H.U., Gray, H.B., Valiente, R. (2004) *Chem. Phys. Lett.* **384**, 190.

Xiong, R.-G., You, X.-Z. (2002) *Inorg. Chem. Commun.* **5**, 677.

Xu, C.N., Watanabe, T., Akiyama, M., Zheng, X.G. (1999) *Appl. Phys. Lett.* **74**, 1236.

Yamamoto, T., Muramatsu, Yu., Lee, B.-L., Kokubo, H., Sasaki, Sh., Hasegawa, M., Yagi, T., Kubota, K. (2003) *Chem. Mater.* **15**, 4384.

Zarkhin, L.S. (2002) *Vysokomol. Soedin., Ser. A Ser. B* **44**, 1550.

Zheng, Zh., Wang, J., Liu, H., Carducci, M.D., Peyghambrian, N., Jabbourb, G.E. (2002) *Acta Crystallogr., Sect. C: Crystal Structure Commun.* **C58**, m50.

Zhu, A., Mio, M.J., Moore, J.S., Drickamer, H.G. (2001) *J. Phys. Chem. B* **105**, 3300.

Zink, J.I. (1978) *Acc. Chem. Res.* **11**, 289.

3 Organic Reactions within Lubricating Layers

3.1 INTRODUCTION

Chapter 3 deals with the chemical changes of lubricating additives and base oils induced by boundary friction. In most mechanical systems (transport, energy production, manufacturing), lubricants are used to reduce friction and wear between moving parts. Lubricants generally consist of mineral, synthetic, or plant (vegetable) oils and contain low concentrations of different additives, including chemical compounds that adsorb on or react with the metallic surface. At present, the following five consumption routes are distinguished for lubricants and greases (Mel'nikov 2005): (1) evaporation; (2) forcing from the friction zone to the reserve zone, followed by centrifugal ejection and spreading/migration in the oil bulk; (3) oxidation; (4) thermal decomposition; and (5) tribochemical reactions. The first two factors are conventionally assigned to physical processes; the last three are assumed to be chemical.

Tribochemistry comprises conventional chemical reactions occurring in the bulk lubricant at the contact zone and mechanically and thermally induced reactions at the metal asperities. The contact between two macroscopic surfaces involves thousands of microasperities. These asperities are small (typical radii of curvature are 10–50 nm). Amonton's law states that friction is proportional to the applied load. Boundary lubrication occurs when there is marked loading (and usually high temperatures) between two rubbing surfaces. Lubricant components react with the contact surface to form lubricant films. The films produce an organic or inorganic thin layer, which reduces wear and friction. Under friction temperature and pressure conditions, additives in the film chemically react with the metal surfaces to form a surface coating. This allows metal-to-metal contact without causing any scuffing or wear. This surface coating acts like a "solid lubricant." Yet, the presence of tribochemical films is not inherently sufficient to protect against wear: efficacy is highly dependent on the chemical and mechanical properties of the film.

Mechanical action at solid surfaces tends to promote chemical reactions and produce surface chemistry, which may be entirely different from the chemistry in static conditions. Frictional work generates energy that is consumed for creation of fresh surfaces and plastic and elastic deformations.

Boundary lubrication is defined by the properties of the surfaces and the lubricants — the properties other than viscosity. As described in Chapter 1, surface friction generates the emission of electrons, photons, phonons, ions, neutral particles, gases (e.g., oxygen), and x-rays. Electron emission is termed the *Kramer effect*. Emission of low-energy electrons and strong warming are particularly important in initiating tribochemical reactions. Therefore, representative examples follow that illustrate effects of electron emission, donor–acceptor interaction, and warming.

3.2 REACTIONS OF LUBRICATING MATERIALS WITH TRIBOEMITTED ELECTRONS

Electron emission occurs when plastic deformation, abrasion, or fatigue cracking disturbs a material surface. Triboelectrons are emitted from freshly formed surfaces. The emission reaches a maximum immediately after mechanical initiation. When mechanical initiation is stopped, the emission decays with time. Strong emission has been observed for both metals and metal oxides. There is strong evidence that the existence of oxides is necessary. The exoelectron emission occurs from a clean, stain-free metallic surface on adsorption of oxygen (Ferrante 1977).

Goldblatt (1971) explained the lubricating properties of polynuclear aromatics by assuming that radical anions are generated at the freshly abraded surface. Low-energy electron (1–4 eV) emission (exoemission) creates positively charged spots on a surface, generally on top of surface asperity. At the same time, the exoemission produces negatively charged radical anions of lubricant components. The positively charged metallic surface attracts these negatively charged radical anions.

In reality, the atmosphere of the tribological system is important. Two usual components of the atmosphere are substantial for boundary lubrication: oxygen and water vapors. Appeldorn and Tao (1968) and Goldblatt (1971) showed that boundary lubrication is not effective in dry argon. In dry argon and in the presence of methylnaphthalene or indene, wear scar diameters are respectively 0.82 or 0.93 nm. Remarkably, these diameters are 0.33 or 0.72 mm, respectively, in dry air and only 0.36 or 0.33 mm, respectively, in wet air.

The following sequence of chemical transformations is obvious:

$$RX + e \rightarrow (RX)^{-}; (RX)^{-} \rightarrow R^{\cdot} + X^{-}; R^{\cdot} + O_2 \rightarrow ROO^{\cdot}; ROO^{\cdot} + H_2O \rightarrow ROOH + HO^{\cdot}; ROOH \rightarrow RO^{\cdot} + HO^{\cdot} \text{ and so on}$$

The radicals formed are involved in further reactions that result in formation of polymers and organometallics. Whereas radical reactions within organic additive mixtures lead to polymeric films, organometallic compounds are understandable as products of interaction between the metallic surface and the radicals. Both polymeric films and organometallic species can protect the rubbing surface from wear. Sometimes, introduction of ready-made metal complexes with organic ligands into lubricating compositions brings a positive effect (Chigarenko et al. 2004; Sulek and Bocho-Janiszewska 2003). Destruction of these complexes originates the polymeric film and leads to liberation of the metal from the complex. The metal is doped into the mating metal surface. This process results in metallurgical changes on the metal surface, making it harder than the steel core of the lubricated device. Importantly, the protective layer formed is continuously renewed.

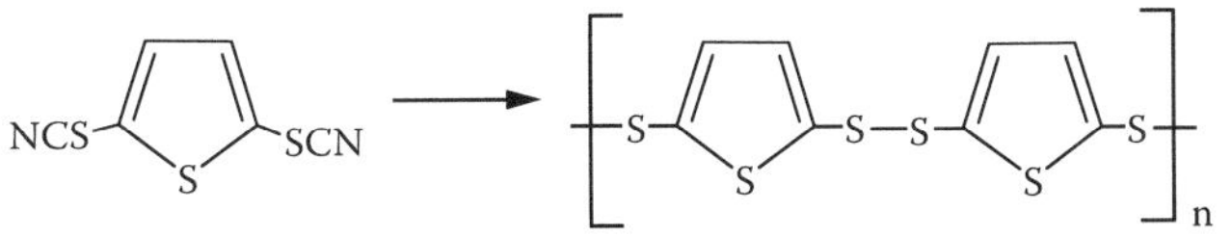

SCHEME 3.1

Many known structural features of lubricant activity become understandable in the framework of the radical ion conception. Thus, it is generally accepted that the extreme pressure performance of disulfides (R—S—S—R) is better than that of monosulfides (thioethers, R—S—R). The difference was simply explained with the radical-ion conception. Monosulfides are reduced less readily than disulfides. Because of the nature of the antibonding orbital hosting the unpaired electron, the sulfur–sulfur bond is elongated seriously in R—S—S—R. Dissociation energy of this elongated bond becomes much smaller than that of the neutral molecule regardless of the aryl or alkyl character of embracing substituents (Antonello et al. 2002). Reductive cleavage of disulfides with generation of active species RS⁻ and RS˙ proceeds readily. Accordingly, disulfides exhibit more efficient load-carrying properties than monosulfides.

Dithiyl radicals seem to be especially attractive because they can form homopolymer films. If one uses a simple compound that is capable of forming dithiyl radical, this can open a way to formulation of a lubrication composition. For example, a poly(disulfide) was obtained as a result of a two-electron transfer to 2,5-di(thiocyanato)thiophene (Scheme 3.1) (Todres 1991).

This linear polysulfide obviously formed according to the sequence 2,5-di(thiocyano)thiophene → potassium 2-mercaptido-5-thiacyanonothiophene → (tristhio)maleic anhydride ⇔ thiophene-2,5-disulfenyl biradical (the diradical valence tautomer) → the depicted linear polymer in which the thiophene rings are connected via disulfide bridges (Scheme 3.2).

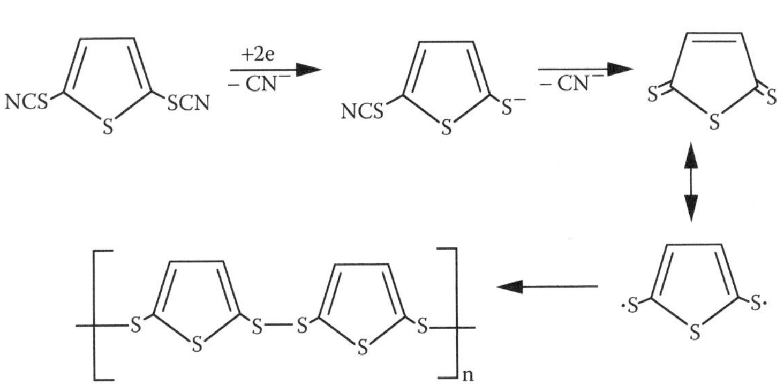

SCHEME 3.2

It was confirmed that trithiomaleic anhydride is unstable and polymerizes just at the moment of formation (see Paulssen et al. 2000 and reference 15 therein). The mentioned reaction provides a good reason to probe such simple compounds as sources of lubricating films for rubbing metallic surfaces.

The work function of the rubbing surfaces and the electron affinity of additives are interconnected on the molecular level. This mechanism has been discussed in terms of tribopolymerization models as a general approach to boundary lubrication (Kajdas 1994, 2001). To evaluate the validity of the radical anion mechanism, two metal systems were investigated, a hard steel ball on a softer steel plate and a hard ball on an aluminum plate. Both metals emit electrons under friction, but aluminum was produced more exoelectrons than steel. With aluminum on steel, the addition of 1% styrene to hexadecane reduced the wear volume of the plate by more than 65%. This effect considerably predominates that of steel on steel. Friction initiates polymerization of styrene, and polymer formation was proven. It was also found that lauryl methacrylate, diallyl phthalate, and vinyl acetate reduced wear in an aluminum pin-on-disk test by 60–80% (Kajdas 1994).

Triboemission also explains steel lubrication by perfluoropolyalkylethers (PFPAEs). PFPAEs possess remarkable properties that make them the lubricant of choice in demanding applications such as magnetic recording media, the aerospace industry, satellite instruments, and high-temperature turbine engines. The physical properties of PFPAEs that enable them to perform lubricating functions in severe environments are their nonflammability, excellent viscosity index, low pour point, and low volatility. Chemically, the unique property of PFPAEs is their stability up to 370°C in an oxidizing and still metal-free environment (Helmick and Jones 1994). Though PFPAEs have excellent thermal stability in a metal-free environment, their stability is significantly decreased to about 180°C when metal alloys and a metal oxide surface are present (Koka and Armatis 1997). This poses a major problem in the practical utility of these lubricants because metals and metal oxides are prevalent in tribological operations. Our consideration concerns the chemical nature of the events in the frictional contact area as well as description of some approaches to circumvent these deteriorating factors.

Let us consider the tribological behavior of one major commercial product of the PFPAE series. This product bears the trade name of PFAE-D; its structure is $CF_3(OCF_2)_x(OCF_2CF_2)_y(OCF_2CF_2CF_2)_zOCF_3$. The molecule contains OCF_2O units that have been attributed to the lower thermal stability of the material compared with other commercial fluids. To gain insight into the decomposition mechanism of the polymer, Matsunuma and coauthors (1966) selected a compound containing five (CF_2O) units as a model for the commercially equivalent fluid. The authors performed a semiempirical molecular orbital calculation comparing optimized structures with the energies of bond breaking. In comparison to the neutral molecule, the formation heat of the corresponding radical anion was markedly lower. Electron attachment to the neutral species loosened the C–O bond. Cleavage of the weakest C–O bond of the radical anion produced the anion and the neutral radical:

$$F(CF_2O)_5CF_3 + e \rightarrow [F(CF_2O)_5CF_3]^- \rightarrow F(CF_2O)_3^- + F(CF_2O)_2CF_2$$

The shortened anion formed still has a weak C–O bond. The degradation of this anion proceeds in a stepwise manner:

$$F(CF_2O)_3^- \rightarrow F(CF_2O)_2^- + CF_2O\uparrow$$

$$F(CF_2O)_2^- \rightarrow CF_3O^- + CF_2O\uparrow$$

Favorable factors for this successive degradation are the low estimated activation energy and the vaporization of the carbonyl fluoride product at room temperature. Strom et al. (1993) observed degradation of PFPAEs during friction tests on magnetic disks using a mass spectrometer. The successive degradation of PFPAEs, which produces carbonyl fluoride, was monitored. This process was called an *unzipping mechanism* (Kasai 1992; Strom et al. 1993).

Zinc dialkyl/diaryl dithiophosphates are widely used as antiwear additives in engine oils to protect heavily loaded mechanism parts from excessive wear. They are also used as antiwear agents in hydraulic fluids. The salts are effective oxidation and corrosion inhibitors; they also act as detergents. During friction, these salts form radical anions that are successfully cleaved (Kajdas et al. 1986):

$$(RO)_2P(S)S-Zn-SP(S)(RO)_2 + e \rightarrow (RO)_2P(S)S^- + [ZnSP(S)S(RO)_2]^{\cdot}$$

Polymers originating from $[ZnSP(S)(RO)_2]^{\cdot}$ radical form lamellar aggregates, which contain Zn, P, S, O, and C but not Fe (Berndt and Jungmann 1983; Sheasby and Rafael 1993). These aggregates transform into zinc sulfide and nonmetallic polymers at high temperatures. Probably, some intimate interaction between the lubricant, and the steel surface forgoes generation of the radicals. This additive takes part in other reactions besides electron transfer. Some of these reactions are considered here, in the section devoted to effects of warming developed during friction. In the conditions of electron transfer, thione-thiol rearrangement of the type P(S)OR $\rightarrow$ P(O)SR takes place (Kajdas et al. 1986). The rearrangement changes the bond polarizability of metal dialkyldithioposphates and enhances the driving forces of all chemical processes described previously.

Triboemission takes place at the asperities of a freshly abraded metallic surface. The surface spots gain a positive charge as a result of triboemission. In other words, fresh surfaces prepared by machining metals are for a short period chemically hyperactive. As shown (Smith and Fort 1958), such surfaces adsorb nonadecanoic acid from cyclohexane solution to a monolayer. Attainment of this coverage is a consequence of a chemical reaction between the fatty acid and the atoms of the fresh metal surface; this is activated by emission of exoelectrons from the surface. This chemical reaction produces a metal soap. Adsorption is a function of the activity of the metal, degree of activation of the machined surface, lifetime of the activated state, and the free-energy requirements of the soap-forming reaction. The adsorbed films are not static. Soap molecules are continually desorbed. The vacancies thus created in the adsorbed layer are filled by diffusion of additional fatty acid to the surface, and then the reaction proceeds *in situ* to form additional molecules of the adsorbed metal soap. The rates of this desorption and adsorbed film replenishment decrease with the age of the metal surface. Kinetics of both these processes and of the initial chemisorption are a function of the surface activation. This idea is substantiated by the fact that the exchange does not take place at a uniform rate across the machined surfaces. At certain immobile sites, exchange is many times more rapid compared with other, less-active areas. The sites on the metal surface where exchange is rapid are identical to the sites from which the low-energy electrons are emitted.

Naturally, when such renewable films acquire a polymeric structure, their lubricating activity is enhanced. Thus, the wear test of a four-ball machine showed that dihydroxydocosanoic acid had a good antiwear property approaching that of zinc dialkyldithiophosphate, the traditional antiwear agent. The order of antiwear properties for the C_{22} acids was 13,14-dihydroxydocosanoic acid > 13- or 14-monohydroxydocosanoic acid > docosanoic acid (no hydroxy substituent). One can see that the antiwear properties of base stock are improved by introducing the hydroxyl group into docosanoic acid. It was revealed by Auger electron or infrared spectroscopy that dihydroxy docosanoic acid formed an oxygen-enriching protective film. Namely, this acid produces the netlike polyester at the expense of the reaction between the carboxyl and hydroxy groups (Hu 2002). It is the tribochemical reaction within the boundary lubrication that is the basis of the improved antiwear activity of dihydroxydocosanoic acids.

Interestingly, 5-allyl-2-methoxyphenol is not polymerized during friction but undergoes transformation into 2-methoxyphenol-5-(methylcarboxylate) under the action of triboemission (Molenda et al. 2003). Seemingly, the carboxylate then forms chelate compounds with the steel surface iron.

Note that, to understand the chemical behavior of lubricant components during boundary lubrication, the concept of triboemission should be examined. The concept is based on the ionization mechanism of lubricants caused by the action of low-energy electrons (1–4 eV). The electrons (exoelectrons) are spontaneously emitted from the fresh surfaces formed during friction. The principal thesis of the model is that lubricant components form radical anions, which are then chemisorbed on the positively charged areas of rubbing surfaces. The model encompasses the following major stages: (1) the low-energy process of electron emission and the creation of positively charged spots; (2) the interaction of emitted electrons with lubricant components and the generation of radical anions, anions, and radicals; (3) the reactions of these radical anions and anions with positively charged metal surfaces, forming films to protect the surface from wear; and (4) the cracking of chemical bonds to produce other radicals. The model explains many lubrication phenomena in which antiwear and extreme-pressure additives are involved. It spurs the design of new additives and lubricating compositions.

3.3 BOUNDARY LUBRICATION AND CHEMISORPTION

Contact and friction within boundary lubrication include the effect of asperity–adhesion forces. These forces can be physical and chemical in the nature. Chemisorption is considered here in this section; physisorption is the subject in connection with warming effects on lubricity and "solvency" of base oils in Sections 3.4 and 3.5, respectively.

Usually, metallic surfaces slide against each other in an oxidative environment (in air). Metal oxides form in this environment. These oxides are Lewis acids and can react with additives during lubrication. It has been shown that the PFPAE molecular chains are prone to undergo the intramolecular disproportionation reaction (Kasai 1992):

$$R-O-CF_2-O-CF_2-OR \rightarrow R-O-CF_3 + R-COF$$

Lewis acids catalyze the disproportionation. It occurs when the successive ether oxygen flanking a difluoromethyylene unit ($-CF_2-$) comes into contact with Lewis acid sites. On wear, some part of the rubbing metallic surface remains covered with the metal oxide. The oxide acts as a Lewis acid coordinating to lone pairs of the ethereal oxygen. Degradation of PFPAE at sliding contacts (with warming) leads to the formation of amorphous carbon and iron fluoride. The latter also acts as a strong Lewis acid (Cheong and Stair 2001). Coordination of this type results not only in fragmentation of the organic molecular chains, but also in generation of molecular fragments possessing a fluorocarbonyl end group. The latter readily contacts with moisture and converts to a strong corrosive mixture of carboxylic and hydrofluoric acids:

$$R-COF + H_2O \rightarrow R-COOH + HF$$

Knowledge of such a process generates an elegant approach to prevent the formation of the corrosive mixture indicated above. Namely, dialkyl amine end groups were introduced into a PFPAE molecule (Kasai and Raman 2002). Ethers of the general formula $R_2N-(CF_2-O-CF_2)_n-NR_2$ contain NR_2 groups, which are stronger as Lewis bases than oxygen in the (CF_2-O-CF_2) fragment. It is the NR_2 group that interacts with the Lewis acid (the metal oxide) instead of the oxygen. The lubricant is resistant to the degradation process, and its working life is significantly longer.

The initial step of coordination between an additive molecule and a metal oxide of the rubbing surface plays an important role in lubricating phenomena. Thus, antiwear properties of organosulfur and organophosphorus compounds in nonpolar synthetic esters were compared by the steel four-ball wear test (Hasegawa et al. 2002). Organophosphorus compounds exhibit excellent antiwear properties, whereas organosulfur derivatives do not display an antiwear effect under the same conditions. The authors explained the results on the basis of the coordination theory. Namely, the stronger bonding between an additive and a transition metal oxide contributes to reduction of wear (if the polarity of the oil medium does not destroy the coordination). The coordination complex further transforms into iron–phosphorus inorganic derivatives.

For instance, tri(cresyl) phosphate (TCP), heated with iron, loses the cresol fragment and gives iron phosphate. Heated in the absence of iron, this TCP does not significantly degrade. With pure iron, a simple iron phosphate forms, and this phosphate is not a lubricant. When iron oxide is present, a polymeric film is formed between cross-linked PO_3 and the iron surface. Tribological reaction of tributyl phosphite [$(BuO)_3P$] with iron oxides leads to formation of a hard polyphosphate glass film (produced by rapid diffusion of PO_x fragments into the oxide) covered by a graphitic carbon layer (Gao et al. 2004). A layered graphitic structure of course produces a lower friction coefficient, particularly when deposited onto a hard polyphosphate glass substrate. This combination of a low-shear-strength material (graphite) deposited onto a hard substrate (a polyphosphate glass) may be the explanation for the efficacy of phosphorus-containing organic compounds as antiwear additives.

Notably, the performance of phosphorus-containing organic additives in a vacuum (in satellite working parts at an oxygen deficit) is not nearly as good as in an air environment (Saba and Forster 2002). Chromium adsorbs TCP less strongly than iron and does not initiate fragmentation of the lubricant. TCP does not lubricate rubbing chromium surfaces (Abdelmaksoud et al. 2002).

The surface of a lubricated metal in air atmosphere exists as an oxide layer. However, naked metal atoms are also created during sliding. When we discuss the adsorption onto the interface of lubricant molecules, two adsorption centers must be taken into account: the metal oxide layer and the freshly opened metal spots. So, lubricant adsorption onto the metal surface in air atmosphere should be considered in its complexity.

Tan and coauthors (2002, 2004) analyzed the interaction between lubricant polar end groups (such as carboxy, hydroxy, and ester) and an aluminum surface on the molecular-orbital level. This allowed prediction of the alcohol-ester friction-reducing effect in the process of aluminum metalworking. (Owing to their excellent resistance to corrosion, superior fatigue resistance, good thermal conductivity, and moderate costs, aluminum alloys have been used widely in industries.) Both alcohols and esters adsorb onto the metal oxide layers by hydrogen bonding. From the results of molecular-orbital calculations, the strength of hydrogen bonds involving alcohol is greater than that involving ester. When alcohol and ester interact with naked aluminum atoms, it is the ester that is sorbed more strongly. So, each of two components has its advantage in interaction with the aluminum oxide and naked aluminum atoms on the surface. Scheme 3.3 illustrates chemisorption of diethyl succinate on a fresh metallic surface (Kajdas and Al-Nozili 2002).

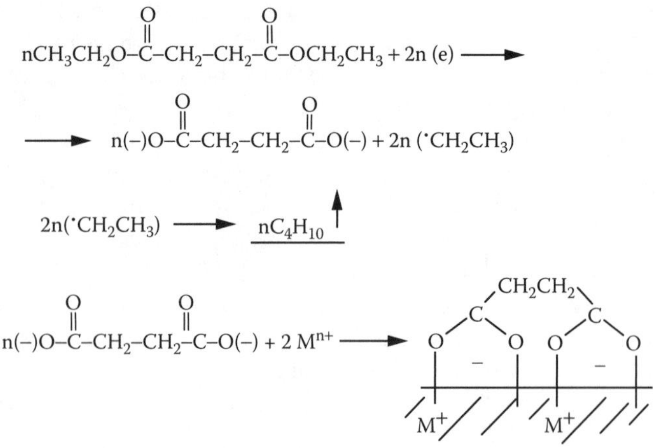

$$nCH_3CH_2O-\overset{\overset{\displaystyle O}{\|}}{C}-CH_2-CH_2-\overset{\overset{\displaystyle O}{\|}}{C}-OCH_2CH_3 + 2n\ (e) \longrightarrow$$

$$\longrightarrow n(-)O-\overset{\overset{\displaystyle O}{\|}}{C}-CH_2-CH_2-\overset{\overset{\displaystyle O}{\|}}{C}-O(-) + 2n\ (^\cdot CH_2CH_3)$$

$$2n(^\cdot CH_2CH_3) \longrightarrow \underline{nC_4H_{10}} \uparrow$$

$$n(-)O-\overset{\overset{\displaystyle O}{\|}}{C}-CH_2-CH_2-\overset{\overset{\displaystyle O}{\|}}{C}-O(-) + 2\ M^{n+} \longrightarrow$$

SCHEME 3.3

On dynamic conditions under the influence of high temperature and electrons emitted from freshly uncovered metal surfaces, saponification of the diester takes place. At the same time, spots appear where metal cations are located. After that, the surface metal cations are chelated by the carboxylic groups. The chemisorption described here starts further tribochemical transformations.

Hydrogen bonding within lubrication layers should also be taken into account. Of the two PFPAEs Z-dol and Z-tetraol, Z-tetraol is more effective. Both have the same molecular mass, 2000, but differ in their end group: $HOCH_2—CF_2O—[CF_2CF_2O]_m—[CF_2O]_n—CF_2CH_2OH$ (Z-dol) and $HOCH_2CH(OH)CH_2O—CH_2—CF_2O—[CF_2CF_2O]_m—[CF_2O]_n—CF_2CH_2OCH_2CH(OH)CH_2OH$ (Z-tetraol). The additional hydroxyl end groups in Z-tetraol increase the possible number of inter- and intramolecular hydrogen bonds, which increases the level of adhesion to the underlying surface and reduces slipping of the lubricant film during friction (Waltman 2004).

However, too strong a connection of the lubricant with surfaces can enforce its discharge. There is some indication that Z-diac commercial lubricant is less efficient than Z-dol and Z-tetraol (Mori et al. 2004). The lubricant Z-diac has the same molecular mass, 2000, but bears carboxylic groups at its ends: $HOOC—CF_2O—[CF_2CF_2O]_m—[CF_2O]_n—CF_2COOH$ (Z-diac).

Zhang et al. (2003) studied rubbing of smooth glass slides covered with fatty acid Langmuir–Blodgett (LB) films against a steel ball. They found that the antiwear behavior of arachidic acid (C_{20}) was better than that of stearic acid (C_{18}) and behenic acid (C_{22}). The fatty chain length is the decisive factor for the quality of LB film. Behenic acid (C_{22}) has the best filming ability, but the tribological behavior of LB film depends on the stiffness as well as the toughness of the molecular chain. The fatty acid molecule consists of the polar carboxylic group and an alkyl radical. Ordered multilayers made of this type of molecule have proper stiffness and toughness, the motion of molecules is more restrained, and the higher shear force during relative motion of the stiff film results in an increase in the friction coefficient. Moreover, the structure of the film changes easily during the friction process because of an increase in strain, so the film has poorer antiwear behavior. By shortening the fatty chain length, LB films can decrease the toughness of the LB film, but the quality of the film will deteriorate. The optimal tribological behavior is reached at balanced stiffness and toughness within the homologous series (see also Adhavaryu 2004a).

LB films are used in many special cases of lubrication. The LB films of nano-particles modified with organic molecules are superior to films of long-chain organic molecules in terms of resisting wear. Zhang et al. (2003) ascribed this to the enhanced load-carrying capacity of the inorganic nanocores in the LB films of nanoparticles modified with organic molecules. During sliding against steel, LB films first transfer onto the counterface during the rubbing process and then undergo a sequence of tribochemical changes, including order transformation and decomposition of alkyl chain. The film transfers and adheres to the counterface tenaciously. LB films on surfaces modified with nanoparticles exhibit better antiwear ability than organic molecules with long-chain films because of the high load-carrying capacity of nanoparticle nanocores. In other words, the nucleus of the nanoparticle plays an important role in increasing the load-carrying capacity and antiwear ability with an increased number of friction cycles.

3.4 WARMING EFFECT ON LUBRICANTS UPON FRICTION

Let us consider now the warming effect on lubricant additives during friction. We discuss the mechanothermal effect produced by the mechanical action itself, but not the trivial heat of chemical reactions that takes place in the friction area. A general model has been developed for the operation of extreme-pressure lubricating additives; it proposes that they thermally decompose at the hot interface (see Lara and Tysoe 1998 and references therein). The *hot interface* means that temperatures around 1000 K can be attained, at least within local temperature pulses (Rusanov 2002). Thermal conductivity of lubricants is less than that of steel. For polymer films, for instance, it is about two orders less (Yamamoto and Takashima 2002). The heat dispersion ability of the rubbing materials is an important moment in boundary lubrication.

X-ray absorption near-edge structure spectroscopy has been used to investigate the chemistry and thickness of thermal and antiwear films generated on steel from oil solutions containing phosphate ester additives (Najman et al. 2002). Diarylphosphates react with steel to form a thermal phosphate film at lower temperatures than triarylphosphates. Substitution of hydrogen for an aryl group in triarylphosphate leads to better wear protection of the metal on tribochemical conditions. For triarylphosphates, a brief period of wear to metal is necessary to initiate the tribochemical reaction between the additive and the metal surface. Once the tribochemical reaction begins, triarylphosphates are able to generate a film of relatively the same thickness and chemistry as diarylphosphates. Najman with colleagues (2002) connected the difference with the following two different reactions of basic iron sites with the triaryl or diaryl derivatives:

$$Fe-OH \text{ (surface)} + OP(OAr)_3 \rightarrow Fe-O-P(O)(OAr)_2 + ArOH$$

$$Fe-OH \text{ (surface)} + OP(OH)(OAr)_2 \rightarrow Fe-O-P(O)(OAr)_2 + H_2O$$

The replacement of an OAr group by an OH in triarylphosphate (to yield diarylphosphate) opens a path to a classic acid–base reaction. Activation energy of this reaction is lower, and it takes place at lower temperatures and more rapidly than the ester interchange (the first of the two reactions).

X-ray absorption studies revealed the formation of iron complexes on milling and warming of $(RO)_2P(S)S-Zn-SP(S)(RO)_2$ mixture with iron oxides. The complexes further lose the alkyl chains and undergo oxidation. The nearest neighbor to the central zinc changes from sulfur to oxygen, followed by Zn–O bond cleavage (Ferrari et al. 2002). As obvious, a manner of the $(RO)_2P(S)S-Zn-SP(S)(RO)_2$ tribochemical decomposition depends not only on electron impact but also on warming during the friction. Thus, under ball-on-disk test conditions, the salts are not totally converted, whereas under the four-ball test (extreme pressure conditions), there are no organic compounds adsorbed on the wear scar surface. The very high temperature in the zone of contacting surface asperities, even reaching 1000°C, is a driving force not only for modification of the surface chemical composition but also for diffusion of Zn, P, and S atoms into the surface (Tuzhynski et al. 2002).

It is very important to keep the integrity of the metal dithiophosphate additive until its interaction with the metal surface. For example, motor oils work on conditions of overheating. The dithiophosphates undergo thermolysis in the critically heated oil well before its interaction with the metal surface to be lubricated. To retard this thermal destruction, adamantane diesters are recommended as an additional additive to motor oils. The adamantane diesters probably form complexes of the van der Waals type with metal dithiophosphates. The tricyclodecane bodies of these diesters shield weak bonds of metal dithiophosphates and keep them from premature cleavage. Having been transported to the metallic surface, the complexes break down because the van der Waals binding becomes secondary to physical sorption. This eventually enhances the thermal stability of motor oil formulations and extends their working lifetime in high-speed motors (Piljavsky et al. 2004).

Lanthanum and neodymium dialkyl dithiophosphates (soluble in oils) show better tribological characteristics than even their zinc analog. X-ray energy dispersion analysis indicated that these metals diffuse into the wear spot surface under friction, and an La- or Nd-rich layer was found. This is supposedly caused by the formation of a boundary film containing neodymium/lanthanum sesquioxide, iron sulfide, sulfate, phosphate, alkyl sulfides, and, importantly, a diffusion layer enriched with rare metal. The layer changes the crystal structure of the friction materials surface, which improves the lubrication performance (Feng, Sun, et al. 2002; Feng, Yan, et al. 2002).

The additive forms a lubricating film on the surface. Physical adsorption of the additive begins immediately at lower temperatures. Natural rubbing of the metallic surfaces increases the oil temperature in the boundary region. An increase in the oil temperature accelerates desorption of the film (Ni et al. 2002). Moreover, the film is continuously worn from the surface under the high load. High load is encountered during extreme-pressure lubrication. Therefore, the resulting thickness of the interfacial lubricating film arises from a balance between the rate of its reactive formation and tribological removal. The nature of the film that is formed depends on the additive used. However, one common event is suggested to occur: the formation of a sublayer after full degradation of the additive.

Meanwhile, the formation of an organic primary film is the very important phase of the tribochemical transformations because it results in adsorption of the film to the metallic surface. In this sense, the structural features of the initial additive can play the key role in lubrication. Thus, zinc dialkyldithiophosphonate (ZnDDP) additives with different types of alkyl groups exhibit differences in both the rates and the by-products of the film formation as well as in the antiwear capabilities of the resulting film. On warming, antiwear film formation is faster for secondary ZnDDPs, is slower for straight-chain primary alkyl ZnDDPs, and is slowest for branched alkyl ZnDDPs. In addition, films derived from secondary alkyl ZnDDP exhibit antiwear properties superior to those of primary alkyl ZnDDPs. All of the alkyl-substituted ZnDDPs react to produce precursors to antiwear films in different ways.

For instance, the Et-ZnDDP system, a model for straight-chain primary alkyl ZnDDPs, produces precursors to the antiwear films through alkyl group transfer from oxygen to sulfur. Such transfer occurs through the elimination of Et_2S. (Large amounts of dialkyl sulfide are usually produced during the formation of films from straight-chain primary alkyl ZnDDPs.) In the case of the iso-Pr-ZnDDP system, a model for secondary alkyl ZnDDPs, the molecule primarily decomposes through olefin elimination. A model for branched primary alkyl ZnDDPs, the iso-Bu-ZnDDP system, also undergoes olefin elimination. However, this is not the main decomposition pathway, and further decomposition forms real antiwear films (Mosey and Woo 2004 and references therein). According to calculations by the B3LYP (density functional theory) Approach, the iso-Pr-substituted system has a lower dissociation barrier in comparison with Et-ZnDDP and iso-ZnDDP (Mosey and Woo 2004).

Macroscopic factors should also be taken into account during consideration of the performance of ZnDDPs. Anticipating development of new technologies, researchers pay attention to limitations on the use of ZnDDPs as additives in boundary lubrication. Lin and So (2003) reported that no protective film was formed on the rubbing surface under high contact pressure, large surface hardness, or low concentration of ZnDDP in paraffin base oil at high temperature. Some limitations were also revealed regarding the rates of recovery and growth of ZnDDP protective film. If ZnDDP concentration was less than 0.5 wt% and it was used above 200°C, then it lost its antiwear properties after several hours of rubbing.

Dialkenyl and dialkyl sulfides are other relevant and important examples of additive degradation between rubbing surfaces. Microsample four-ball friction and wear tests were conducted to evaluate the mechanism of dialkenyl sulfides as additives in liquid paraffin (Han et al. 2002). No polymer was detected after friction at a relatively low load. The friction-reducing and antiwear capacities in this case were mainly attributed to chemisorption of the starting additive onto the metal surface. Under medium load, friction did lead to high molecular weight polymer. In this case, the friction-reducing and antiwear capacities were attributed to friction-induced polymerization. Under a relatively high load, the friction polymer experienced degradation. The inorganic boundary film composed of ferrous sulfide and sulfate contributed to improve significantly the friction-reducing and antiwear behavior in the last case.

After all the initial transformations at the boundary lubrication, dialkylsulfides form alkyl thiolate fragments distributed on the rubbing metal parts. The alkyl thiolate fragments eventually decompose to deposit carbon and sulfur onto the surface. At high temperatures, the carbon and sulfur diffuse into the bulk of the metal, forming metal sulfide and carbide. Namely, the metal sulfide film is deposited onto a carbon-enriched metal surface. According to Kaltchev and coworkers (2001), such film structure arises because carbon diffuses into iron more rapidly than sulfur. In this case, the interface consists of the ferrous sulfide on the carbon–iron underlayer. The interfacial temperature is 1400 ± 100 K for both dimethyl and diethyl disulfides as starting additives. Importantly, this temperature is close to the melting point of ferrous sulfide (1460 K).

Consequently, when sulfur-containing additives are used, films containing a ferrous sulfide layer are deposited onto a carbided iron contact surface. As is known, the hardness of iron increases considerably with the addition of a small amount of carbon. The friction coefficient μ of a thin film with shear strength S_F when deposited on a surface of hardness H_S is given by $\mu = S_F/H_S$ (Bowden and Tabor 1964). Because H_s for the iron-containing inclusions of iron carbide is higher than that for the "pure" iron, this leads to substantially decrease the friction coefficient. That is why the μ values are relatively low for sulfur-containing additives.

Tribological performance of borated alkyl dithiocarbamate $[(C_8H_{17})_2NC(=S)SCH_2CH_2OB(OC_8H_{17})_2]$ also results in thermal destruction. Iron-surface analysis indicated the formation of a protective film of borate and sulfur/nitrogen organic derivatives adsorbed on the rubbed surface. However, the elemental sulfur reacts with the metal surface and forms iron sulfate and sulfide, $FeSO_4$ and FeS_2, respectively. Boron nitride is also formed. Such complicated protective film forming during the sliding process significantly enhances the load-carrying performance and the anti-wear properties of the lubricating mixture as against zinc dialkyldithiophosphate (Huang et al. 2002), which must be replaced by other, zinc- and phosphorus-free, additives in the near future. This is the reason for the intense comparison of new compounds to zinc dialkyldithiophosphate.

The same (enhanced) tribological behavior was described for additives depicted in Scheme 3.4, namely, (2-thiocarbonylbenzothiazole)-3-methylene dibutyl borate (W. Huang, Dong, et al. 2001; W. Huang, Li, et al. 2001; W. Huang et al. 2002), N-alkyl/alkylene imidazoline borates (Gao et al. 2002), and aminoethyl alkylborate esters (Yao et al. 2002).

The third (last) member of Scheme 3.4 represents a lubricant with improved hydrolytic stability. In general, borate esters are nonvolatile and relatively nontoxic and have a pleasant odor. However, a serious drawback that has restricted the use of borate esters in lubricant oils is their susceptibility to hydrolysis. Hydrolysis liberates oil-insoluble and abrasive boric acid. Formation of the nitrogen-to-boron coordination bond (the same as that in the structure under consideration) inhibits hydrolytic attack on the boron–oxygen bonds of the esters (Yao et al. 2002). Regarding iron sulfate formation on steel, sulfur additives were shown to decompose at 100°C (prior to friction). The elemental sulfur is oxidized to iron sulfate on the surface. Oxygen is easily provided by iron oxides on the surface or from dissolved oxygen in the oil. Friction leads to much higher local temperatures, resulting in partial depletion of the oxide layer. This involves the bare metal in the interaction, with the sulfur giving rise to the formation of pyrite, FeS_2. The further warming effect (at more than 800°C) causes pyrite to decompose to pyrrohotite, FeS (Najman et al. 2003). In mechanochemistry, some indirect phenomena should be taken into account to understand the effects of additives and to design syntheses of new ones.

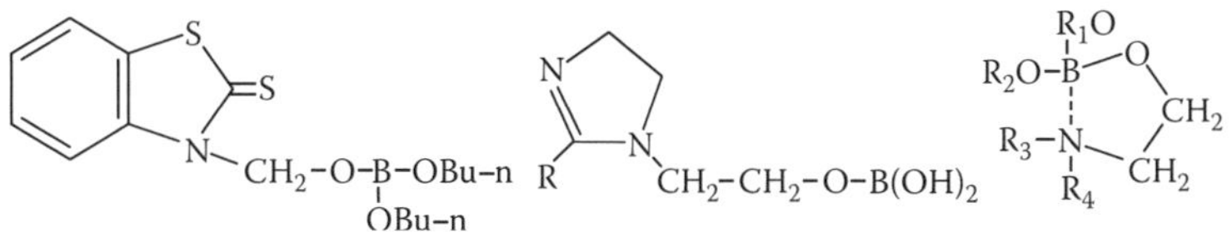

SCHEME 3.4

Concerning relative roles of tribochemical and thermochemical processes, some authors (e.g., Tysoe and others 1995) assumed that tribochemical reactions are simply provoked by the high temperature induced during metal–metal contact (because of plastic deformation, etc.). Other authors (e.g., Martin 1999) believed that there is something unique about the high-pressure and high-shear situation existing in tribological systems. Piras and coworkers (2002) distinguished thermochemical and tribochemical reactions as developing in two different mechanisms with different kinetics; the tribochemical reaction seems to be faster than the thermal reaction. Such a point of view is a middle one and therefore is nearer to the reality.

The necessity of co-joint consideration of thermochemical and thermal reactions is especially important in the case of vapor phase lubrication. The development of high-efficiency engines operating at extremely high temperatures requires lubrication schemes compatible with operating temperatures in excess of 500°C. Vapor phase lubrication is used as a method for the lubrication of high-temperature engine components. Within such a scheme, vaporized lubricants are delivered continuously in a carrier gas to the hot sliding surfaces of engine components. The vapor phase lubricant reacts to deposit a thin solid lubricating film on these surfaces. At present, typical vapor phase lubricants are phosphorus-containing organics, including arylphosphates such as tri(cresyl)phosphate $[(CH_3C_6H_4O)_3PO]$ and alkylphosphates such as tri(butyl)phosphate $[(C_4H_9O)_3PO]$. Importantly, $(CH_3C_6H_4O)_3PO$ is more effective than $(C_4H_9O)_3PO$ for the deposition of lubricious films (Ren and Gellman 2000). According to the Auger spectra, the cresyl derivative is capable of depositing substantial amounts of carbon and some phosphorus onto the iron surface. At the same time, the butyl derivative deposits only phosphorus (Sung and Gellman 2002a, b).

Tri(butyl)phosphate decomposes via C—O bond cleavage to produce butyl groups $[CH_3CH_2CH_2CH_2—]$ connected with the surface, which further decompose via β-hydride elimination, producing 1-butene $(CH_3CH_2CHCH_2)$ and H_2. These volatile gases are desorbed, and no appreciable carbon deposits onto the surface. In contrast, tri(cresyl)phosphate generates the surface-bonded $(CH_3C_6H_4O—)$ groups by O–P bond scission and then the surface-bonded $(CH_3C_6H_4—)$ groups after C—O bond scission. These $(CH_3C_6H_4O—)$ groups cannot undergo β-hydride elimination to desorb from the surface simply because they do not have β-CH bonds. Therefore, $(CH_3C_6H_4—)$ groups transform on the surface into unstable species such as benzyne (cyclohexadienyne). Because they are extremely reactive, these species do not have enough time to desorb from the surface before their further fragmentation and reaction with the iron surface. This leads to high rates of carbon deposition and to the above-mentioned differences in performance of alkylphosphates and arylphosphates as vapor-phase lubricants.

3.5 "SOLVENCY" AND REACTIVITY OF BASE OILS

In many cases, the antifriction performance of additives is strongly related to the solvency of the oil medium used. This is the case for molybdenum dialkyldithiocarbamate (MoDTC). Base oils of low aromatic content provide relatively poor solvation for MoDTC, and friction reduction is optimized (Shea and Stipanovic 2002). The organomolybdenum molecule preferentially adsorbs to metal surfaces under poor solvation. As a low-percentage additive, MoDTC is completely dissolved in the base fluid (solvent). It is the solvation that makes the additive solubility possible. The important point is that the solvation forces should be weaker than the additive affinity to the surface. As an additive, MoDTC very rapidly forms a thick film. During friction, MoDTC gives rise to molybdenum sulfide and oxide. The oxide is not good at reduction of friction, whereas the sulfide is excellent. So, the average friction of MoDTC is good. Wear control is outstanding (Yamaguchi et al. 2005).

When MoDTC is incorporated into a lubricant formulation, many other additives are typically present. These molecules include ZnDDP (considered in Section 3.4), to control wear, and surface active detergents and dispersants, which help neutralize acidity and prevent deposits. During normal use of engine oil, MoDTC is largely solvated in the bulk rather than adsorbed onto engine surfaces. In the oil solution bulk, ligand exchange reactions with other additives can take place. The chemical nature of the base oil, the operating temperature of the lubricant, and engine surface temperatures also can influence the actual process of MoDTC adsorption onto engine parts. Once adsorbed, MoDTC is transformed into a friction-reducing agent under tribological conditions. Presumably, the Mo-S core in MoDTC is decomposed to form MoS_2, which has been shown to be a superior friction reducer (Yamamoto and Gondo 1994). This layer possesses good sliding properties and is easily shearable because of its layered or sheetlike structure. It acts as a very good lubricant between the surfaces. However, owing to its easily shearable nature, there is a possibility that at high load its rate of formation is lower than the rate of removal. In practice, this impairs the high-load performance (Unnikrishnan et al. 2003).

Note that the organic ligands attached to the molybdenum-sulfur core are crucial because they allow the compound to gain solubility in lubricant base oils. Without long alkyl groups, the molybdenum-sulfur core would be absolutely insoluble in oil, thereby limiting its use as an additive. In principle, MoDTC may be envisioned as a precursor that eventually transforms into the active friction-lowering compound molybdenum disulfide on the engine surface under tribological contact (see also Muraki, Aoyangi, and Sakaguchi 2002). Similarly, molybdenum complex with *N*-isonicotinyl-*N*'- (4-methoxyphenyl)thiosemicarbazide liberates oxygen from the methoxyphenyl group. The oxygen further interacts with a metal to form additional oxides alongside sulfides of molybdenum and iron on the iron sliding surfaces (Rastogi and Yadav 2004).

The friction-reducing effect of MoDTC deteriorates over time in actual engine applications, and this activity reduction typically occurs before all of the MoDTC additive is depleted from the engine oil. As Shea and Stepanovic (2002) pointed out, if ligand exchange leads to the formation of a new molybdenum complex with much longer alkyl ligand, the solubility of the complex might increase sufficiently. As a result, it no longer adsorbs to the metal surface, thereby inhibiting friction reduction. Alternatively, ligand exchange, which reduces the alkyl ligand chain below a critical solubility threshold, might cause precipitation of the corresponding molybdenum complex.

The polarity of base oil also affects the ligand exchange rates because the more polar oil promotes faster exchange. The addition of detergent was shown to reduce ligand exchange rates. The added detergent interacts strongly with polar MoDTC (Shea and Stepanovic 2002). This effectively lowers the probability for ligand exchange.

Consequently, understanding the events in the oil solution phase (before the additive adsorbs onto the engine surface) is very important for correct engine oil formulations to improve engine efficiency. At this point, chemical reactions resulting in solubility of the surface itself deserve especial consideration. Ceramic materials of the Si_3N_4 type attract widespread interest as metal substitutes in automotive and jet engines, including general-purpose Army tactical vehicles. In particular, these materials are used for running-in fired engines. The low thermal expansion coefficient, high hardness sustainable at elevated temperatures, and other mechanical properties enable the material to exhibit excellent resistance to thermal shock, abrasion, and creep. Displacing lead-containing alloys, they do not pollute the environment. At the same time, contemporary fuels contain petroleum and lower alcohol constituents. Hibi and Enomoto (1997) examined the mechanochemical reaction of Si_3N_4 with lower alcohols. This reaction actively promoted both the elution of silicon nitride into alcohols and dehydration condensation of the alcohols. This resulted in the formation of silica gel and higher hydrocarbons, alkanes, and alkenes. It is obvious that the reaction considered here is directly connected with wearability of engine working parts.

Concerning antifriction additives for ceramic materials, lubrication by ionic liquids should be pointed out as a novel approach. Ionic liquids have a number of excellent properties, including negligible volatility, nonflammability, high thermal stability, and controlled miscibility with organic compounds. The pour points for the ionic liquids are about −50°C, and the liquids are excellent candidates for vacuum lubrication. They also have good potential for lubrication of steel–steel, steel–aluminum, and steel–copper contacts, exhibiting friction reduction, antiwear properties, and heavy load capacity (Omotowa et al. 2004 and references therein).

Ye et al. (2002) studied lubricity of 1-methyl-3-octylimidazolium tetrafluoroborate during sliding of dysprosium-sialon ceramics against Si_3N_4. This sialon ceramic can be represented with the empirical formula $Dy_{0.33}Si_{9.3}Al_{2.7}O_{1.7}N_{14.3}$. The ionic liquid was used as such, with no base oil. 1-Methyl-3-octylimidazolium tetrafluoroborate exhibits a very low friction coefficient under both low and high load. It demonstrates superior tribological properties to perfluoropolyethers or glycerol. The imidazolium is very easily adsorbed on the frictional pair contact surface because it has the polar structure depicted on Scheme 3.5.

During friction, ionic liquids give rise to boron sesquioxide and iron nitrides as well as ferrous fluoride. These compounds (formed from ionic liquids) are good solid lubricants that effectively reduce friction and wear at a steel–steel contact (Liu et al. 2002). Notably, ionic liquids are ideal for lubrication of devices and apparatus working in cosmic conditions.

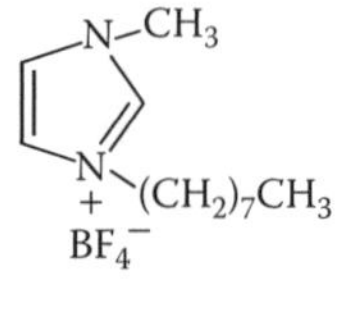

SCHEME 3.5

Another modern and important approach consists of formulations based on phosphorus- and nitrogen-containing modifications of rapeseed oils dissolved in nonmodified rapeseed oils. In accordance with the similarity principle, the solubility of such additives is better in the rapeseed oils than in the mineral oils. According to Fang and others (2002), these formulations are effective enough and provide better antiwear and friction-reducing properties in rapeseed oil than in mineral oil. The mechanism of their activity remains the same (the formation of iron phosphide/ nitride films). However, one supplemental property is achieved: biodegradability. Both base oil and additive are biodegradable. This property is extremely important at present.

Modification of vegetable base oil "in bulk" seems to be interesting. Such oils contain glyceryl tricarboxylates (stearate, palmitate, laurate). Polar functional groups in the triglycerol molecule interact with the metallic surfaces under high and sliding contact. Increasing the polar functionality of vegetable oil might have a positive impact on wear protection resulting from stronger adsorption on a metal surface. The polar groups attach to the metal, whereas the nonpolar alkyl chain ends form a molecular layer separating the rubbing surfaces. For instance, a positive effect was achieved with soybean oil (Adhvaryu et al. 2004b). This oil was epoxidized and converted into the corresponding dihydroxylated product; then, the hydroxyl groups were esterified with hexanoic anhydride.

Another principal example concerns palm kernel oil (Yunus and colleagues 2004). The methoxy groups were initially introduced in the α-CH positions of carboxylates, and then the ethereal esters were treated with tris(methylol)propane. The product of transesterification had improved lubrication characteristics. It demonstrated superior oxidative stability as well as a lower pour point than the original palm oil products. In addition, palm-based synthetic lubricants retain biodegradability. The best antiwear and antifriction characteristics were found with modified samples of palm kernel oil that contained up to 15% tris(methylol)propyl diesters. Chain branching, adhesive, solvency, and other effects considered in this chapter can explain the lubrication results of such intricate chemical modification. Meanwhile, the commercial prospects of the approach are unclear if one takes into account the huge volume of materials to be treated by the method considered.

There are other attempts in this direction. Fiszer et al. (2003) proposed a procedure for transesterification of rapeseed oil and utilization of the products as components of mixtures with conventional base oils. The first stage was the methanolysis of rapeseed oil in the presence of alkaline catalysts (K_2CO_3, CH_3ONa, NaOH). The obtained methyl esters were subjected to transesterification with the higher alcohols trimethylolpropane, 2-ethylhexanol, and triethyleneglycol. The resulting esters were used to prepare semisynthetic oils. Namely, the esters were added to mineral oil at a volume of 25–50%. The mixed oils had better lubricity and a higher viscosity index. The authors indicated that rather unfavorable oxidative stability of semisynthetic oils containing the esters is considerably improved by application of oxidation inhibitors.

It is known that environmental pollution caused by conventional mineral lubricants is significant because of the low biodegradability of these oils and their toxicity. The environmental acceptability of lubricants has become of concern worldwide. Vegetable oils are attractive because of their biodegradability and low ecotoxicity.

They are cheap and available from renewable sources. Traditional rapeseed oil is usually preferentially selected as the base stock with a view of acquiring a rational balance between performance and cost. The German government issued "Blue Angel Regulations," as legal and permissive standards for industries. Korff and Cristiano (2000) enumerated standards for the lubricants and greases. The corresponding formulations must be metal free and must not contain chlorine because of toxicity. At the same time, sulfur or phosphorus can be introduced because each has low toxicity. For instance, zinc dialkyldithiophosphate [ZnDDP] forms destruction products. These products are unhealthy for humans. Therefore, ZnDDP would eventually be prohibited, although it has received the widest use. It is multifunctional additive for lubricants with high load-carrying capacity as well as antiwear, friction-reducing, and antioxidation properties (Spikes 2004).

The formulation based on ZnDDP and rapeseed oil is disadvantageous in view of the tendency of the oil to autooxidate. Rapeseed oil is the triglyceride of unsaturated acids with a peroxide value of 309 ppm. This value exceeds 2000 ppm after 17 months of storage. The autooxidation leads to the formation of peroxides ROOH. The peroxides produce an antagonistic effect on the antiwear properties of [ZnDDP](Minami and Mitsumune 2002). The reaction of peroxides with [ZnDDP] gives the corresponding disulfide:

$$Zn[DDTP] + ROOH \rightarrow ZnO + ROH + DDTP$$

Because the antiwear properties of disulfides are inferior to those of $[Zn(DTP)_2]$, the antagonistic effect of peroxides is understandable.

Li and coauthors (Li, Zhang, Ren, and Wang 2002; Li et al. 2003) used zinc-free dialkyldithiophosphate ester additives in rapeseed oil. The formulation provides good load-carrying capacity similar to that of zinc dialkyldithiophosphate. In the process of rubbing, tribochemical reactions occur between the additives and the metal surface. Phosphorus reacts with the metal, producing inorganic metal salts; sulfur is not detected on the surface. Li and colleagues (Li, Zhang, Ren, Liu, and Fu. 2002) also tested S-(1H-benzotriazole-1-yl) methyl N,N-dialkyldithiocarbamates (TADTC on Scheme 3.6) as additives in rapeseed oil. These additives possessed excellent load-carrying abilities and high thermal stability but had no antiwear and friction-reducing properties. Such a result points to a very subtle relation between polarity of an additive and of an oil.

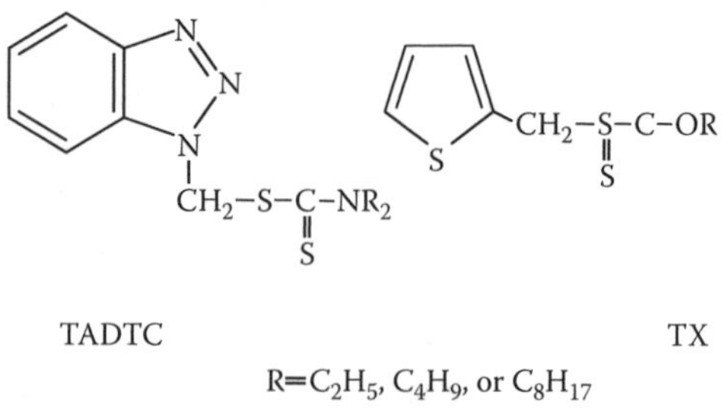

TADTC   TX

$$R=C_2H_5, C_4H_9, \text{ or } C_8H_{17}$$

SCHEME 3.6

The polarity of rapeseed oil can lead to competitive adsorption on metal surfaces between the additives and the base oil. Benzotriazole thiocarbamates have much lower polarity than the base oil. When added to the oil, the additives cannot generate boundary lubricating layers on the friction surface but reduce the tenacity of the lubricating film formed by the base oil because of the competitive adsorption. So, the antiwear and friction-reducing abilities of the oil were decreased with the addition of these compounds. Under heavy load, the additives decomposed during the sliding process. The elemental sulfur therein reacted with the metal surface and formed a protective film. According to x-ray photoelectron spectra, the film contained $FeSO_4$ or $Fe_2(SO_4)_3$. The parts containing the elemental nitrogen were adsorbed on the metal surface by the N atom only (Li, Zhang, Ren, Liu, and Fu. 2002). In total, improved load-carrying capacity was observed in the presence of such a protective film.

As an example of a good balance between rapeseed oil and additive polarity, the tribological behavior of triazole dithiocarbamates (TADTC) and [S-(2H-thiophen-2-yl)]-methylalkyl xanthates (TXs) should be mentioned; see Scheme 3.6 (Gong et al. 2002).

Some discussion should be made regarding tribochemistry of base oils. Triglycerides of fatty acids that form a base of a vegetable oil liberate the fatty acids within the gap between sliding metallic surfaces. The acids are considered a classical boundary lubrication improver. As shown for sunflower oil, addition of free fatty acid to the oil results in little effect if the acids are similar to those already present. Among C-18 acids of sunflower triglycerides, only addition of their minor component, stearic acid, dramatically improved antifriction and antiwear properties of this oil in the temperature interval from 50 to 150°C (Fox et al. 2004).

Influenced by the processes occurring under friction conditions, hydrocarbon oils can also undergo reactions that lead to the formation of chemically active compounds. In this regard, the tribochemistry of hexadecane should be considered. This hydrocarbon is widely used as a model reference fluid in testing additives for both metal and ceramic surfaces. Using a ball-on-disk tester (bearing steel as the tester material), tribochemical changes of normal hexadecane were studied in detail (Kajdas et al. 2003; Makowska et al. 2002). During the friction process, the hydrocarbons (as bulk lubricants) transformed into ketones, aldehydes, alcohols, carboxylic acids, and esters. These products appeared as a result of a reaction connected with electron triboemission (Makowska et al. 2004) through formation of superoxide and hydroxy radical as considered in Section 3.2. All the products are potential triboactive compounds that can react with iron atoms on the steel surface. Indeed, wear tracks contain iron carboxylates or complexes and, near the steel, iron carbide.

McGuiggan (2004) measured the friction and adhesion between a fluorocarbon monolayer-coated surface against a hydrocarbon monolayer-coated surface. The friction was lower than the friction between a hydrocarbon monolayer against a hydrocarbon monolayer and a fluorocarbon monolayer against a fluorocarbon monolayer. A fluorocarbon chain is stiffer than a hydrocarbon chain. Because the lateral adhesion within the hydrocarbon monolayer is greater than the adhesion between the hydrocarbon/fluorocarbon monolayer, little interdigitation occurs. All these factors reduce the interlayer friction. Such a peculiarity is industrially important: By coating the interior wall of the extruder, the fluorine-containing compounds can reduce the shear stress at the extruder surface.

Consequently, tribochemical reactions of bulk oils can, in their turn, make a contribution to boundary lubrication. These reactions are specifically traced to the disrupted surface caused by the sliding contact generating surface-active sites, promoting reactions that otherwise might not occur. Such changes in the bulk oil lead to deterioration of lubricating oils.

3.6 CHEMICAL ORIGINS OF ADDITIVE SYNERGISM–ANTAGONISM

The combined requirements of solubility, adhesion, antiwear, and extreme-pressure antifriction are usually addressed in lubricant formulations. They are essential for sustaining protection of equipment under a variety of operating conditions. Typically, additive function passes from the solution to the surface, thermally decomposing there and then reacting to form a solid protective layer. Antiwear additives operate under mild conditions, while extreme-pressure additives usually have a higher activation threshold and are used in lubricating environments in which both the operating temperatures and the loads are high. The resulting film fills surface asperities, thereby minimizing contact of sliding surfaces and reducing events such as friction, repeated adhesion, welding of opposing (metallic) surfaces, and surface wear.

Regarding the chemical composition of the resulting film, phosphate esters are regarded as antiwear additives; organosulfur compounds are considered extreme-pressure elements of oil formulations. Under boundary lubrication of steel sliding parts, they react to generate layers of iron sulfide and sulfate. Both antiwear and extreme-pressure lubricant additives containing phosphorus and sulfur are polar molecules that will eventually meet each other at the metal surface. When they are introduced into base oil co-jointly, competition for surface sites can lead to the inefficiency of one or more of them. (In addition to this, additive–additive interactions in the oil solution can also occur, leading to possible synergistic or antagonistic effects.) This deserves to be illustrated with some representative examples.

Najman and colleagues (2004) monitored properties of tribochemical films generated from oil solutions comprised of two additives containing phosphorus and sulfur, for instance, diphenyl phosphate and dialkyldithiocarbamate (named Irgalube 349 by Ciba-Geigy). Tests were performed under both antiwear and extreme-pressure conditions. The study found that the chemistry of the tribofilms formed was clearly dominated by the phosphate ester additives. Formation of the glasslike phosphor-containing product apparently blocked surface sites and hindered the sulfur–additive reaction with the metal. In the presence of diphenyl phosphate, Irgalube 349 was oxidized to sulfate in the bulk of the film. The proportion of sulfur was significantly reduced compared to the films generated by the sulfur-containing additive in the absence of the phosphate ester. Wear protection was determined completely by the action of phosphorus. Sulfur had little role, if any, in further protecting the metallic surface.

Of course, complex action of an additive mixture is a goal of a smart formulation, especially if some synergistic effect is attainable. The work of Evdokimov et al. (2002) provided a good example. Basic oil (Tap-15, Russia) was supplied with minimal amounts of high-density polyethylene, poly(ethyl siloxane), and chromium(III) acetylacetonate. The composition reduced the friction coefficient and mass wear by about 10% compared with traditional hydrocarbon lubrication media. It is underlined that chromium acetylacetonate is not a complex capable of reducing the friction coefficient, and poly(ethyl siloxane) is not one of the better lubricants for heavy-duty friction units.

Evdokimov and coworkers (2002) suggested the following explanation for such unexpected improvement in antifriction/antiwear properties. Hydrolysis takes place with formation of chromium(III) hydroxide in local zones because of warming during friction. This is the reaction between chromium(III) acetylacetonate with air moisture and lubricant adsorbed on the surface of the steel. Chromium hydroxide [$Cr(OH)_3$] rapidly decomposes into chromium sesquioxide (Cr_2O_3) and water in the same conditions. Cr_2O_3 is a good, highly disperse abrasive. During friction, it smoothes the roughness of steel surfaces, increasing the surface finish class, and grinds off superfluous oxide film layers. The most stable oxide films particularly include Fe_3O_4 and $FeO \cdot Cr_2O_3$. The mixed oxide $FeO \cdot Cr_2O_3$ is probably formed on the surface of the steel by chromium(III) acetylacetonate in the form of a film, which protects the surface from corrosion and wear. However, particles from wear consisting of fragments of the initial oxide layers of the steel and Cr_2O_3 particles could be scraped off and could destroy the steel surface during friction. As a constituent of the formulation, the poly(ethyl siloxane) molecules adapt to the surface structure of the oxide wear particles and are firmly sorbed. The oxide eventually turns into the micellar form, which is probably not dangerous to the worn friction surface.

As for the high-density polyethylene molecules and the oil base molecules, they undergo tribo-oxidative decomposition under the effect of oxygen, which is always present in the oil. New compounds with polar functional groups (COOH, OH, COOR, etc.) are obtained. In this way, surfactant molecules form. On origination, the surfactants create mono- and multimolecular adsorption layers that ensure easy slipping in boundary friction. Chromium acetylacetonate and poly(ethyl siloxane) thus play a special role in the proposed mechanism. The first compound, by decomposing in the friction zone, forms the abrasive chromium sesquioxide, and when adsorbed, the second compound converts wear particles of the sesquioxide film on the steel surface into the micellar form, which prevents further wear of the surface and favors the formation of polymolecular adsorption layers on them.

When MoDTC is used together with zinc dialkyldithiophosphonate (ZnDDP) as the principal antiwear additive, a synergistic effect on friction and wear characteristics is observed. ZnDDP undergoes destruction at first. Once the rubbing surface is covered sufficiently with the products from ZnDDP, the formation of MoS_2 from MoDTC is promoted and better retained on it (Bec et al. 2004; McQueen et al. 2005; Muraki and Wada 2002). This results in a drastic decline in the coefficient of friction. In the presence of overbased calcium borate as a detergent, the MoDTP/ZnDDP mixture acts even more effectively. The friction coefficient is lowered from 0.07 to 0.05. For such a formulation, MoS_2 sheets have also been found as a result of the tribochemical reaction. Because it is embedded in a single-phase calcium and zinc borophosphate glass, MoS_2 is perfectly oriented. It is the orientation effect that lowers the friction coefficient in this case (Martin et al. 2003). There are data that

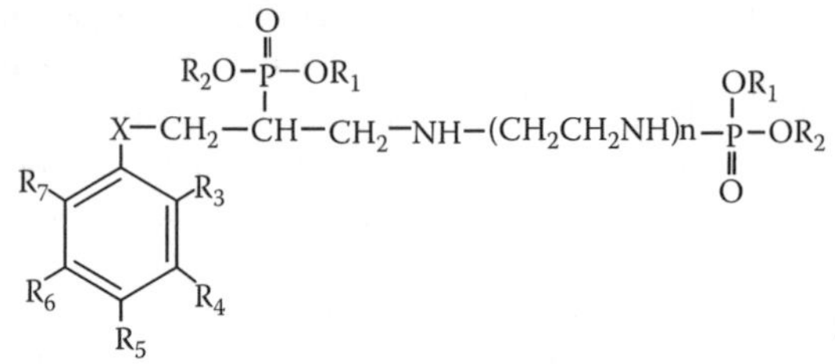

$$R_2O-\overset{\overset{\displaystyle O}{\|}}{\underset{|}{P}}-OR_1$$

$$X-CH_2-\overset{|}{CH}-CH_2-NH-(CH_2CH_2NH)n-\overset{\overset{\displaystyle OR_1}{}}{\underset{\displaystyle O}{P}}-OR_2$$

(aromatic ring with substituents R_7, R_3, R_6, R_4, R_5)

SCHEME 3.7

boron-containing compounds facilitate the decomposition of metal dithiophosphates and, correspondingly, formation of tribofilms (Zhang, Yamaguchi, Kasrai, and Bancroft 2004).

One interesting approach consists of hanging different moieties responsible for the activities needed on the same carcass. The organic molecules depicted on Scheme 3.7 were designed as multifunctional lubricants. According to a Taiwan patent claim (Ye et al. 2000), the lubricants contain an aromatic amine or phenol moiety for its antioxidant ability, an aliphatic amine moiety for dispersive physisorption, and a phosphoric ester moiety for extreme pressure-resistant stability. The long-chain alkyl introduced to the molecule renders good solubility to the lubricant additive in mineral oils (Scheme 3.7).

3.7 MOLECULAR MECHANISMS OF DRY-SLIDING LUBRICATION

Life sometimes poses situations when liquid lubrication is totally prohibited. For instance, agricultural machines work in high-dust conditions; therefore, oil greases are the source for enhanced wear. In such an environment, dry lubrication is a choice. Among organic dry lubricants, polymers are frequently used. The chemical mechanism of their action should also be the subject of consideration.

In the case of polymers, the friction energy is expendable for cleavage of bonds at the points of physical contact. The polymeric radicals from ruptured bonds recombine to form layers with lower molecular masses at the surface. The future decomposition takes place at the expense of such a newly formed layer with the formation of tear products. In reviewing the literature data, Zhang and He (2004) denoted that the molecular segments of the freshly ruptured polymers adhered to the metal surface under the action of van der Waals forces. Then, a lubrication adhesion layer was formed on the surface of indentation. The free radicals reacted with the metal oxide surface and produced a metal–oxide organic complex. This complex acted as a lubricant to retard wearing. However, there are more subtle regularities concerning the polymer protective properties.

Krasnov and coworkers (1996) studied the structure of the surface layers formed by ultrahigh molecular weight polyethylene or polycaproamide under friction. The authors observed self-organization of the polymer materials on the friction surface. This process included active mass transfer from the surface to the underlying layers and orientation of the polymer chains in the direction of friction.

A nanometer-thick surface polymeric layer was formed by the particles with lower surface energy.

Gent and Pulford (1978) compared *cis*-polyisoprene and *cis*-polybutadiene as protectors of steel from wear. The *cis*-polybutadiene was a better protector. It forms more reactive radicals than *cis*-polyisoprene does. The more active radicals react mainly with the parent polymer, forming a new one with higher molecular mass. The less-active radicals react with the parent polymer slowly, meanwhile diffusing to react with steel directly. It is clear that the high-polymer protector acts longer than the protector of a lower molecular mass.

Silicon-containing polymers form layers in the friction area. With such layers, strain more proportionately spreads on the contact zone; shear stress is transferred from the materials of contact surfaces into the material of the dividing layer. During friction, especially in severe regimes, silicon-containing polymers are homolytically cleaved. As a result, solid silicate particles appear to be confined in a shell of a milder polymeric (oligomeric) material. The silicate particles are distinguished by high thermal stability. This avoids disruption of the structured adsorption layers during frictional warming. The lubricating material formed as a result of the reaction between sodium silicate and phenol-formaldehyde resin (resol) is an example (Zlotnikov and Volnyanko 2001). This system is a control base for the highest load capacity allowed for frictional blocks (up to 40 MPa). It is also characterized with the very low coefficient of friction (up to 0.01). At such high pressures, any lubricant solidifies and stays in contact.

Since the beginning of the use of perfluoropolyethers like Z-dol (see Section 3.3), such polyethers with a variety of end groups have been synthesized and examined in an effort to strengthen the interaction (bonding) between the lubricant molecules and the magnetic disk surface. The lubricant of choice for the disk application was Z-dol. Chemical transformations of perfluoropolyethers on friction were considered in this chapter. One remaining point not yet discussed is the nature of the disk surface—the surface that sorbs the polymer lubricant. The magnetic layers of hard storage disks are coated with a thin layer (5–20 nm) of sputter-deposited carbon. This is needed to protect the magnetic layer from corrosion and from the abrasive impact of the head. A thin layer (~1 nm) of Z-dol is applied over the carbon cover. With the lubricant layer 1-nm thick, the interaction between the lubricant molecules and the carbon cover has become the critical factor in disk lubrication performance. According to Kasai (2002), in reference to a private communication of Dr. Benjamin DeKoven a significantly larger fraction of Z-dol is deposited if exposition of the disks to the lubricant vapor immediately follows the carbon sputtering process without prior exposure to air. If disks were exposed to air and then brought to the vapor lubricant deposition stage, the bonded fraction decreased in proportion to the duration of air exposure. The problem is about competition between air oxygen and terminal Z-dol hydroxyl groups for the active spots of the freshly deposited carbon overcoat. Significantly, no bonding occurred when freshly prepared disks were exposed to perfluoropolyether Z-15 (no functional end groups) without prior exposure to air (Kasai 2002).

Krasnov and coauthors (1997) considered relationships among the nature of the repeating unit, the structure of the macromolecular chain, and the frictional behavior of the polymer material. The main recommendations were (1) to use linear or cross-linked polymers containing aromatic (heteroaromatic) rings in the backbone; (2) to diminish the number of reactive groups remaining nonfitted in the polymer chain; (3) to prepare linear polymers with maximal molecular masses; and (4) to use crystalline or liquid crystalline polymers as materials with the most ordered physical structure.

Another prospective class of dry lubricants is represented by fullerenes and their derivatives. As molecules, fullerenes are huge hollow balls with an outer shell composed totally of fused five- and six-member rings. The antifriction behavior of these compounds depends on their application conditions. It was found (Drexler 1992) that, in the force field of anisotropic compression, C_{60} fullerene is transformed into known carbon phases (diamond, graphite) and into some metastable crystalline and amorphous modifications. During shear strain, C_{60} fullerene generates carbon radicals formed on destruction of the C_{60} molecules (Dubinskaya 1999).

At the same time, there are recent examples of lubricating by ball-shaped fullerene derivatives that keep their integrity. The fullerene–itaconic acid copolymer was synthesized by radical polymerization with C_{60}/C_{70} fullerene and itaconic acid. Itaconic acid is propylene-2,3-dicarboxylic acid $[CH_2\!\!=\!\!C(COOH)CH_2COOH]$. The copolymer has the ideal spherical shape; its average diameter is about 48 nm. The copolymer has good load-carrying and antiwear capacities (Guan and Shen 2002). L. Huang and colleagues (2002) claimed that the star-shaped C_{60}-perfluoro-1-octane-sulfonate film has a lower friction force than the star-shaped C_{60}-polystyrene film, which is expected to be a promising lubricant. The authors found that the surface of the C_{60}-perfluoroctane sulfonate thin film possesses lower surface energy. A molecular chain with lower surface energy gives a smoother surface with lower friction force.

Ginzburg and Tochil'nikov (2002), Miura, Kamiya, and Sasaki (2003), Sasaki and Miura (2004), Ren et al. (2004), and Okita and coauthors (2003) considered fullerenes as molecular rolls (similar to ball-thrust bearings) in the friction area. A fullerene film is expected to be a good lubricant because of the nearly spherical shape, low surface strength, and high robustness. Because they are chemically active, fullerene films can modify the superficial layers of counter solids.

Fullerene is an effective scavenger of free organic radicals. Each C_{60} fullerene molecule can add up to six macroradical chains (Hirsh 1998). A novel fullerene–styrene–acrylic acid copolymer was synthesized via radical polymerization. The particles formed had a 36-nm average diameter. The physical structure of the fullerene copolymer is nanometer-size tiny spheres may be described as follows: The core is very hard fullerene, and the shell is styrene–acrylic acid copolymer long-chain moiety. This moiety is relatively soft but elastic. The fullerene copolymer is nanometer-size tiny spheres with a core-shell structure. Penetrating into rubbing surfaces and depositing there, such a polymer acts as a solid lubricant (Jiang et al. 2003).

However, the high costs of the fullerene preparations hinder their practical applications. To circumvent this crucial obstacle, Kireenko and coauthors (2002) paid attention to the waste product in the fullerene manufacture, the fullerene soot. Used as a wear and friction protector, the soot was effective enough. A combination of pressure with shearing action in zones of tribocontact assisted in the formation of supplemental molecules of fullerene from the unfinished fragments contained in the soot. As a result, the fullerene concentration was enough to build up a protective layer on friction surfaces.

Metallofullerenes as dry lubricants should be mentioned as intriguing objects. Fullerenes themselves possess an aromatic character and are stable. However, fullerenes lose aromaticity on electron capture (Sternfeld and Rabinovitz 2002). Usually, aromaticity is associated with stability. Electron capture, as seen in Chapter 2, is one of the lubrication results. Metallation also charges neutral fullerenes negatively, with the metal atom encapsulated at the center of a fullerene shell. Kang and Hwang (2004) theoretically predicted that potassium-fullerene nanoball bearings might be more rigid and applicable than their nonmetallic counterparts. Future experiments should control this theoretical prediction.

3.8 CONCLUSION

The interaction mechanisms between lubricating additives and metal surfaces mainly include physical adsorption, chemisorption, and surface chemical reactions. Boundary lubrication generates emission of electrons, photons, phonons, ions, neutral particles, gases (e.g., oxygen), and x-rays. All emanations play roles in the transformations of additives within boundary lubrication. Chapter 3 is a part of the treatise devoted to organic mechanochemistry. Naturally, it intentionally scrutinizes those of the phenomena that could receive chemical interpretation. Especially, the analysis of films originating from lubricating additives on the rubbing surfaces is crucial for better understanding of the additive action.

Certainly, lubrication phenomena are complicated. Therefore, we are forced to abstract some crucial transformation for their scrutiny. Such an approach is usual in analysis of phenomena, even complicated ones. Accordingly, electron transfer, donor–acceptor interactions, solubility–adsorptivity relationships, and warming effects are considered separately although they represent different constituents of lubricating as a whole. It is worthwhile to underline that lubrication is a widespread strategy to reduce friction. However, it suffers from some limitations in the boundary regime, where the thickness of a lubricating spacer is only several molecular layers. Fluids have a general tendency to attain a solidlike structure when squeezed between two solid surfaces. This confinement-induced solidification is responsible for microscopic stick-slip motion when the confining surfaces are sheared past each other (Jagla 2002 and references therein). In the boundary regime, lubricants composed of branch chain molecules usually perform better than ones formed by linear chain molecules. The antiscuffing activity of branched additives (e.g., unsymmetrical dialkyl disulfides; Kirichenko et al. 2003) is also higher than that of linear ones. As a rule, the branched additive does not readily arrange in a solidlike structure and allows smooth shearing of the surface in a state of very low friction. In addition, fluids with branched-chain molecules have typically higher bulk viscosity than fluids with linear chain molecules. This justifies the experimentally observed (Heuberger and colleagues 1998 and references cited there) negative correlation existing between bulk viscosity and friction coefficient under boundary lubrication conditions. Meanwhile, molecularly thin layers of branched substances give rise to more disordered structures than those of linear ones.

For this reason, the pressure to squeeze the lubricant out decreases when disorder occurs in the lubricant film (Sivebaek and colleagues 2004).

To optimize the balance between low wear and low friction, such lubricant formulations are meritorious if their viscosity is sufficient to generate hydrodynamic or elastohydrodynamic oil films that separate the machine's interacting surfaces, but not too high to induce excessive viscous drag looses. In cases of high-load friction, the squeezing-out phenomenon should also be taken into account during the composition of the lubricating oils. The chapter shows that both macroscopic and molecular approaches should be applied co-jointly, especially regarding new lubricants for new applications. In addition to metalworking, micromechanical devices and high-density magnetic storage media are application fields in which subtle features may affect the system performance in a significant way.

The impetus for new, improved lubricant additives comes from numerous sources. Governmental and regulatory requirements demand lower toxicity. New engine developments, such as the ceramic diesel, are on the horizon, presenting opportunities for antiwear additives that can function at very high operating temperatures. Space technology and other advanced transportation needs present new challenges to the industry. Of course, there will always be a need for low product costs and ease of production. At this point, silicon-containing hydrocarbons represent a relatively new class of liquid lubricants with great potential for use in space mechanisms. As an example, the additive 6-88-134 can be denoted $n\text{-}C_8H_{17}Si[CH_2CH_2CH_2Si(n\text{-}C_{12}H_{25})_3]_3$ (Jones et al. 2004). Silahydrocarbons possess unique antiwear properties, high viscosity, very low volatility, and the ability to solubilize conventional additives.

An emphasis is made on proper ashless additives. Farng (2003) gave the following reasons for such prognosis: More automobiles are equipped with catalytic converters. These converters decrease their catalytic efficiency in the presence of phosphorus derived from zinc dithiophosphates. Therefore, a strong need exists for engine oils with lower phosphorus content. This creates a need in alternative antiwear/extreme-pressure additives and antioxidant/anticorrosion additives. The main task will consist of partial replacement of zinc dithiophosphates, which possess all of these functional properties.

In this connection, attention is drawn to nitrogen-containing heterocyclic compounds. Many works indicated that the excellent properties of those compounds could be attributed to their compact and stable structures (He et al. 2002; Huang, Hou, Zhang, and Dong. 2004; Huang, Dong, Li, and Hou 2004; Huang, Dong, Wu, and Zhang 2004). They can reduce friction and wear and increase the load-carrying capacity of lubricating oil, and they possess anticorrosion, antirust, and antioxidation properties. Even such simple nitrogen-containing compounds as N-alkyl morpholines display low friction and wear coefficients in liquid paraffin base oil at a concentration of 0.5 wt%. All the characteristics are better than those of ZnDDP. The worn surface lubricated by ZnDDP alone contains many corrosive pits. On the worn surface lubricated with alkylated morpholine, very few corrosive pits were found. The additives scavenge active sulfur to provide corrosion inhibition. They form stable adsorption films through the lone pairs of electrons on the nitrogen atoms and act as good ligands with strong coordination capacity (Shui et al. 2002, 2003).

The use of metallic antiwear/extreme-pressure additives should be diminished to prevent environmental deterioration. Heavy metals are considered pollutants, and their presence is no longer welcomed in the environment. Given equal performance and costs, ashless antiwear additives will be preferred for many future lubricants. In the future, the lubricant additive business will continue to grow and will need unconventional antiwear/extreme-pressure additives. Possible new markets include biodegradable lubricants, advanced transportation lubricants, as well as lubricants for robotics, ceramics, and devices that work in space. Traditional markets in engine oils; automatic transmission fluids; and marine, aviation, gear, hydraulic, circulating oils, metalworking, and other industrial lubricants are also expanding, although moderately. Advanced antiwear additives with environmentally friendly features, excellent stability, and unique performance properties will be the additives of choice for increasingly demanding lubricant applications.

All of these topics are reflected in Chapter 3. According to data from the United Nations, world countries spend about 30% of generated electricity for overcoming friction forces in machines and process equipment. A significant portion of gross domestic products of industrial countries is wasted every year because of friction and wear problems. Because of friction, the United States annually loses up to 2% of its gross national product. Tribology experts estimated that these costs could be reduced 25–30%, if already-available findings were used in practice. In 10–15% of cases, it would not demand an additional major invest-ment (Luzhnov 2001). Chapter 3 centers attention on the typical new formulations, explaining the grounds for their useful activity with the intent of helping devel-opment in this engineering field.

REFERENCES

Abdelmaksoud, M., Bender, J.W., Krim, J. (2002) *Tribol. Lett.* **13**, 179.

Adhvaryu, A., Erhan, S.Z., Perez, J.M. (2004a) *J. Agric. Food Chem.* **52**, 6456.

Adhvaryu, A., Erhan, S.Z., Perez, J.M. (2004b) *Wear* **257**, 359.

Antonello, S., Benassi, R., Gavioli, G., Taddei, F., Maran, F. (2002) *J. Am. Chem. Soc.* **124**, 7529.

Appeldorn, Y.K., Tao, F.F. (1968) *Wear* **12**, 117.

Bec, S., Tonck, A., Georges, J.M., Roper, G.W. (2004) *Tribol. Lett.* **17**, 797.

Berndt, H., Jungmann, S. (1983) *Schmier-techn.* **14**, 306.

Bowden, F.P., Tabor, D. *The Friction and Lubrication of Solids* (Oxford University Press, London, 1964).

Cheong, Ch.U.A., Stair, P.C. (2001) *Tribol. Lett.* **10**, 117.

Chigarenko, G.G., Ponomarenko, A.G., Burlov, A.S., Garnovskii, A.D. (2004) *Russ. Pat.* 02238302.

Drexler, R.K. *Nanosystems, Molecular Machinery. Manufacturing and Computation* (Wiley, New York, 1992).

Dubinskaya, A.M. (1999) *Usp. Khim.* **68**, 708.

Evdokimov, Yu.A., Kornev, V.I., Bulgarevich, S.B., Vilenskii, V.M. (2002) *Khim. Tekhnol. Topliv Masel*, **2**, 23.

Fang, J.-H., Chen, B-Sh., Zhang, B., Huang, W-J. (2002) *Synth. Lubr.* **19**, 141.

Farng, L.O. (2003) *Chem. Ind. (Dekker)* **90** *(Lubr. Additives),* 223.

Feng, Yu-J., Sun, L.-X., Sun, X.-J., Hu, Ya., Cai, W.-M. (2002) *Harbin Gongue Daxue Xuebao* **34**, 324.

Feng, Yu-J., Yan, T.-A., Sun, X.-J., Sun, L.-X., Cai, W.-M. (2002) *Zhongguo Xitu Xuebao* **20**, 207.

Ferrante, J. (1977) *ASLE Trans.* **20**, 328.

Ferrari, E.S., Roberts, K.J., Adams, D. (2002) *Wear* **253**, 759.

Fiszer, S., Szalajko, U., Bieg, T., Klomfas, Ja., Niemiec, P. (2003) *Pol. J. Appl. Chem.* **47**, 307.

Fox, N.J., Tyrer, B., Stachowiak, G.M. (2004) *Tribol. Lett.* **16**, 275.

Gao, F., Furlong, O., Kotvis, P.V., Tysoe, W.T. (2004) *Langmuir* **20**, 7557.

Gao, Y., Jing, Y., Zhang, Zh., Chen, G., Xue, Q. (2002) *Wear* **253**, 576.

Gent, A.N., Pulford, C.T.R. (1978) *Wear* **49**, 135.

Ginzburg, B.M., Tochil'nikov, D.G. (2002) *Prob. Mashinostroeniya Nadyozhnosti Mashin*, **2**, 60.

Goldblatt, I.L. (1971) *Ind. Eng. Chem., Prod. Res. Div.* **10**, 270.

Gong, Q.-Ye, Yu, L.-G., Ye, Ch.-F. (2002) *Wear* **253**, 558.

Guan, W., Shen, Ch. (2002) *Caliao Baohu* **35**, 15.

Han, N., Shui, L., Liu, W.-M., Xue, Q.-J., Sun, Yu-Sh. (2002) *Mocaxue Xuebao* **22**, 49.

Hasegawa, T., Hirao, K., Memita, M., Minami, I. (2002) *Toraiborjisuto* **47**, 198.

He, Zh., Rao, W., Ren, T., Liu, W., Xue, Q. (2002) *Tribol. Lett.* **13**, 87.

Helmick, L.S., Jones, R.W. (1994) *Lubr. Eng.* **50**, 449.

Heuberger, M., Drummond, C., Israelachvili, J.N. (1998) *J. Phys. Chem. B* **102**, 5038.

Hibi, Y., Enomoto, Y. (1997) *J. Mater. Sci. Lett.* **16**, 316.

Hirsh, A. In: *Topics in Current Chemistry, vol. 199: Fullerenes and Related Structures*. Edited by Hirsh, A. (Springer, Berlin, Germany, 1998 p. 1).

Hu, Zh. (2002) *Runhua Yu Mifeng*, **1**, 38.

Huang, L., Fan, Sh., Wei, F., Zhao, X-Sh., Xiao, J.-X., Zhu, B.-Ya. (2002) *Chin. J. Polym. Sci.* **20**, 397.

Huang, W., Dong, J., Li, F., Chen, G., Chen, B. (2001) *Lubr. Sci.* **14**, 73.

Huang, W., Li, F., Chen, B., Dong, J. (2001) *Shiyou Lianzhi Yu Huagong* **32**, 54.

Huang, W., Hou, B., Zhang, P., Dong, J. (2004a) *Wear* **256**, 1106.

Huang, W., Dong, J., Li, J., Hou, B. (2004b) *Tribol. Lett.* **17**, 199.

Huang, W., Dong, J., Wu, G., Zhang, Ch. (2004c) *Tribol. Int.* **37**, 71.

Huang, W., Tan, Y., Dong, J., Chen, B. (2002) *Tribol. Int.* **35**, 787.

Jagla, E.A. (2002) *Phys. Rev. Lett.* **88**, 245504/1.

Jiang, G.-Ch., Guan, W.-Ch., Zheng, Q.-X. (2003) *Huaxue Yu Shengwu Gongcheng* **20**, 15.

Jones, W.R., Jansen, M.J., Gschwender, L.J., Snyder, C.E., Sharma, S.K., Predmore, R.E., Dube, M.J. (2004) *Synth. Lubr.* **20**, 303.

Kajdas, Cz. (1994) *Lubr. Sci.* **6**, 203.

Kajdas, Cz. (2001) *Tribol. Ser.* **39**, 233.

Kajdas, Cz., Al-Nozili, M. (2002) *Tribologia* **33**, 861.

Kajdas, Cz., Makowska, M., Gradkowski, M. (2003) *Lubr. Sci.* **15**, 329.

Kajdas, Cz., Tuemmler, R., von Ardenne, H., Schwartz, W. (1986) *Z. Ind. (Leipzig)* **115**, 107.

Kaltchev, M., Kotvis, P.V., Blunt, T.J., Lara, J., Tysoe, W.T. (2001) *Tribol. Lett.* **10**, 45.

Kang, J.W., Hwang, H.J. (2004) *Nanotechnology* **15**, 614.

Kasai, P.H. (1992) *Macromolecules* **25**, 6791.

Kasai, P.H. (2002) *Tribol. Lett.* **13**, 155.

Kasai, P.H., Raman, V. (2002) *Tribol. Lett.* **12**, 117.

Kireenko, O.F., Ginzburg, B.M., Bulatov, V.P. (2002) *Trenie Iznos* **23**, 304.

Kirichenko, G.N., Khanov, V.Kh., Ibraghimov, A.G., Glazunova, V.I., Kirichenko, V.Yu., Dzhemilev, U.M. (2003) *Neftekhimiya* **43**, 468.

Koka, R., Armatis, F. (1997) *Tribol. Trans.* **40**, 63.

Korff, J., Coistiano, A. (2000) *NLGI Spokesman* **64**, No. 8, 22.

Krasnov, A.P., Gribova, I.A., Chumaevskaya, A.N. (1997) *Trenie Iznos* **18**, 258.

Krasnov, A.P., Makina, L.B., Panov, S.Yu., Mit', V.A. (1996) *Trenie Iznos* **17**, 371.

Lara, J., Tysoe, W.T. (1998) *Langmuir* **14**, 307.

Li, J., Rao, W., Ren, T., Fu, X., Lui, W. (2003) *Synth. Lubr.* **20**, 151.

Li, J., Zhang, Ya., Ren, T., Liu, W., Fu, X. (2002) *Wear* **253**, 720.

Li, J., Zhang, Ya., Ren, T., Wang, D. (2002) *Synth. Lubr.* **19**, 99.

Lin, Y.C., So, H. (2003) *Tibol. Int.* **37**, 25.

Liu, W., Ye, Ch., Gong, Q., Wang, H., Wang, P. (2002) *Tribol. Lett.* **13**, 81.

Luzhnov, Yu. M. (2001) *Tyazhyoloe Mashinostroenie*, No. 4, 3.

Makowska, M., Kajdas, Cz., Gradkowski, M. (2002) *Tribol. Lett.* **13**, 65.

Makowska, M., Kajdas, Cz., Gradkowski, M. (2004) *Lubr. Sci.* **16**, 101.

Martin, J.M. (1999) *Tribol. Lett.* **6**, 1.

Martin, J.M., Grossiord, C., Varlot, K., Vacher, B., Le Mogne, Th., Yamada, Ya. (2003) *Lubr. Sci.* **15**, 119.

Matsunuma, S., Toshimasa, M., Kataoka, H. (1996) *Tribol. Trans.* **39**, 380.

McGuiggan, P.M. (2004) *J. Adhes.* **80**, 395.

McQueen, J.C., Gao, H., Black, E.D., Gandopadhyay, A.G., Jensen, R.K. (2005) *Tribol. Int.* **38**, 289.

Mel'nikov, V.G. (2005) *Zashchita Metall.* **41**, 168.

Minami, I., Mitsumune, Sh. (2002) *Tribol. Lett.* **13**, 95.

Miura, K., Kamiya, S., Sasaki, N. (2003) *Phys. Rev. Lett.* **90**, 055509/1.

Molenda, Ja., Gradkowski, M., Kajdas, Cz. (2003) *Tribology* **34**, 93.

Mori, S., Cong, P., Nanao, H., Numata, T. (2004) *Tribol. Lett.* **17**, 317.

Mosey, N.J., Woo, T.K. (2004) *J. Phys. Chem. A* **108**, 6001.

Muraki, M., Aoyanagi, M., Sakaguchi, K. (2002) *Int. J. Appl. Mech. Eng.* **7**, 397.

Muraki, M., Wada, H. (2002) *Tribol. Int.* **35**, 857.

Najman, M.N., Kasrai, M., Bancroft, G.M. (2003) *Tribol. Lett.* **14**, 225.

Najman, M.N., Kasrai, M., Bancroft, G.M. (2004) *Wear* **257**, 32.

Najman, M.N., Kasrai, M., Bancroft, G.M., Miller, A. (2002) *Tribol. Lett.* **13**, 209.

Ni, Sh.-Ch., Kuo, P.-L., Lin, J.-F. (2002) *Wear* **253**, 862.

Okita, S., Matsumuro, A., Miura, K. (2003) *Thin Solid Films* **443**, 66.

Omotowa, B.A., Phillips, B.S., Zabinski, J.S., Sreeve, J.M. (2004) *Inorg. Chem.* **43**, 5466.

Paulssen, H., Haitjema, H., Van Asselt, R., Mylle, P., Adriaensens, P., Gelan, J., Vanderzande, I. (2000) *Polymer* **41**, 3121.

Piljavsky, V.S., Golovko, L.V., Brjuzgin, A.R., Khilchevsky, A.I. (2004) *Cataliz Neftekhimia*, No. 12, 32.

Piras, F.M., Rossi, A., Spencer, N.D. (2002) *Langmuir* **18**, 6606.

Rastogi, R.B., Yadav, M. (2004) *Indian J. Chem. Technol.* **11**, 317.

Ren, D., Gellman, A.J. (2000) *Tribol. Trans.* **43**, 480.

Ren, S., Yang, Sh., Zhao, Ya. (2004) *Langmuir* **20**, 3601.

Rusanov, A.I. (2002) *Zh. Obshcheii Khim.* **72**, 353.

Saba, C.S., Forster, N.H. (2002) *Tribol. Lett.* **12**, 135.

Sasaki, N., Miura, K. (2004) *Jpn. J. Appl. Phys., Part 1* **43**, 4486.

Shea, T., Stipanovic, A.J. (2002) *Tribol. Lett.* **12**, 13.

Sheasby, J.S., Rafael, Z.N. (1993) *Tribol. Trans.* **36**, 399.

Shiu, L., Han, N., Sun, Y., Liu, W., Xue, Q. (2002) *Soc. Automotive Eng., Special Pub.* **SP-1722** *(Lubricants)*, 175.

Shiu, L., Han, N., Sun, Y., Liu, W., Xue, Q. (2003) *Mocaxue Xuebao* **23**, 316.

Sivebaek, I.M., Samoilov, V.N., Persson, B.N.J. (2004) *Tribol. Lett.* **16**, 195.

Smith, H., Fort, T. (1958) *J. Phys. Chem.* **62**, 519.

Spikes, H. (2004) *Tribol. Lett.* **17**, 469.

Sternfeld, T., Rabinovitz, M. (2002) *Khim. beYisra'el* **10**, 3.

Strom, B.D., Bogy, D.B., Walmsley, R.G., Brandt, J., Singh, B.C. (1993) *Wear* **168**, 31.

Sulek, M.W., Bocho-Janiszewska, A. (2003) *Tribol. Lett.* **15**, 301.

Sung, D., Gellman, A.J. (2002a) *Tribol. Int.* **35**, 579.

Sung, D., Gellman, A.J. (2002b) *Tribol. Lett.* **13**, 9.

Tan, Y., Huang, W., Wang, X. (2002) *Tribol. Int.* **35**, 381.

Tan, Y., Huang, W., Wang, X. (2004) *Tribol. Int.* **37**, 447.

Todres, Z.V. (1991) *Electrochim. Acta* **36**, 2171.

Tuzhynski, W., Molenda, Ja., Makowska, M. (2002) *Tribol. Lett.* **13**, 103.

Tysoe, W.T., Surerus, K., Lara, J., Blunt, T.J., Kotvis, P.V. (1995) *Tribol. Lett.* **1**, 39.

Unnikrishnan, R., Christopher, J., Jain, M.C., Martin, V., Srivastava, S.P. (2003) *TriboTest*
    9, 285.

Waltman, R.J. (2004) *Langmuir* **20**, 3166.

Yamaguchi, E., Roby, S., Yeh, S. (2005) *Tribol. Trans.* **48**, 57.

Yamamoto, Y., Gondo, S. (1994) *Tribol. Trans.* **37**, 182.

Yamamoto, Y., Takashima, T. (2002) *Wear* **253**, 820.

Yao, J.B., Wang, Q.L., Chen, S.Q., Sun, J.Z., Dong, J.X. (2002) *Lubr. Sci.* **14**, 415.

Ye, Ch., Liu, W., Chen, Yu., Ou, Zh. (2002) *Wear* **253**, 579.

Ye, M.-R., Guo, B.-L., Huang, Yu.-H., Lin, G.-H. (2000) *Taiwan Pat.* TW 414806.

Yunus, R., Fakhru'l-Razi, A., Ooi, T.L., Iyuke, S.E., Perez, J.M. (2004) *Eur. J. Lipid Sci.*
    Technol. **106**, 52.

Zhang, P., Xue, Q., Du, Z., Zhang, Zh. (2003) *Wear* **254**, 959.

Zhang, S.-W., He, R.-Ya. (2004) *J. Materials Sci.* **39**, 5625.

Zhang, Z., Yamaguchi, E.S., Kasrai, M., Bancroft, G.M. (2004) *Tribol. Trans.* **47**, 527.

Zlotnikov, I.I., Volnyanko, E.N. (2001) *Trenie Iznos* **22**, 689.

4 Mechanically Induced Organic Reactions

4.1 INTRODUCTION

Chapter 4 provides principal examples of mechanically-induced organic reactions. In addition to parameters routinely considered, such as reaction temperature or choice of solvent, emerging mechanical methods of activation also are important, although not frequently taken into account. Cutting, crushing, drilling, grinding, comminution, kneading, friction, shearing, and sliding represent such methods. Shock wave and ultrasonic effects are also considered in this chapter if comparison to mechanical activation is relevant.

Certainly, pressure in the range of 0.1 to 2 GPa strongly influences the rate and equilibrium position of many chemical reactions. Pressure leads to changes in the activation and reaction volumes, in packing manner, and in degree of electrostriction. These pressure effects are out of the scope of this chapter. *Organic Synthesis at High Pressure* by Acheson and Matsumoto (1991) and the collected reviews by van Eldik and Klaerner (2003) are addressed to those interested in the topic.

The mechanical methods mentioned for activation or organic reactions form a major theme of Chapter 4 and are pursued in detail. These methods are becoming a new technique in organic synthesis, especially for large-scale realization. They constitute a way of modifying the conditions in which organic reactions take place. Accordingly, this chapter underlines the advantages of the mechanically-induced reactions regarding those in solutions as follows: shorter reaction time, lower reaction temperature, less workup, no need for solvents, higher or comparable yields, and the possibility to merge different synthetic steps in one-pot synthesis.

From the entire variety of mechanochemical organic syntheses, only methodically significant and scientifically principal examples can be considered because of limited book volume. The material here wholly or partly obeys the following criteria:

- The reactions given must demonstrate important features and conditions of mechanochemical activation.
- The procedures chosen must have definite and tangible advantages.
- Yields of the products must be sufficient.
- The products themselves must be of practical interest or contribute to future practical applications.
- The mechanochemical methods included are sometimes the only ones available for a certain outcome.

This chapter details achievements in mechanochemical organic syntheses not only for their scientific and practical merits, but also for the aesthetic appeal of the examples chosen and the intrinsic beauty of the methods that have emerged. In the best case, these examples obey all the stated requirements. However, some reactions only partly meet the requirements. A goal of this chapter also is to describe general approaches to mechanochemical methodology of organic synthesis more than to provide a detailed account of each example published. The usefulness and general outlines of this methodology define the list of reactions considered in Chapter 4.

4.2 MECHANOCHEMICALLY INITIATED POLYMERIZATION, DEPOLYMERIZATION, AND MECHANOLYSIS

4.2.1 POLYMERIZATION

One of the most-used mechanically induced polymerization processes is so-called reactive extrusion. Solid monomers pass through an extruder within the turbulent flow. To the solids, this flow transmits high pressure combined with shear stress. An example is polymerization of crystalline pentabromobenzyl acrylate with simultaneous grafting of the polymer onto magnesium hydroxide filler (Gutman and Bobovitch 1994). This filler was used as a smoke-consuming agent in combustion of plastics. Polymerization was carried out in a double-screw extruder at high temperature. The activated surface of the inorganic filler behaved as an additional initiator of mechanopolymerization. High temperature, pressure, and shear stress co-jointly led to the fracturing and amorphization of monomer crystalline regions. This was accompanied by the formation of free radicals that acted as polymerization initiators. Further polymerization took place on the filler surface. This improved adhesion between the filler and the polymer. The magnesium hydroxide surface was involved in growth of the polymer chain so that some chemical bonds formed between the polymer and the filler surface. Reactive extrusion in its application to the monomer–filler system produced plastics with reduced flammability and improved mechanical and thermal properties. In addition, the filler replaced part of the organic material. This diminished the toxicity of plastics' production and cost.

Stabilization of poly(vinylchloride) in elastic deformation with a metal stearate is the next example of the process just considered. When ground with poly(vinylchloride), barium, or lead, stearate is effectively distributed within the polymer sample. This enhances the interaction between the stabilizer and chemically active fragments of the polymer and thus protects the polymer against thermal destruction (Akhmetkhanov et al. 2004).

Polymers can also be made by vibromilling of some monomers with steel balls. No initiators are needed. Such polymerization is initiated under the action of the electronic stream developed by mechanoemission under the vibratory milling. Mechanoemission of exoelectrons is known as the Kramer effect. The effect was described in Chapters 2 and 3.

On vibratory milling, acryl and methacryl amides give anion radicals, which are key species in the reaction (Simonescu et al. 1983):

$$CH_2{=}CRCONH_2 + e \rightarrow (CH_2{=}CRCONH_2)^{-\cdot} \ \{R{=}H, CH_3\}$$

$$(CH_2{=}CRCONH_2)^{-\cdot} + CH_2{=}CRCONH_2 \rightarrow {}^-CH_2{-}CR(CONH_2){-}CH_2{-}CR{\cdot}{-}CONH_2$$

Further growth of the polymeric chain proceeds in the usual manner. Compared to polymeric materials obtained by conventional methods, the mechanochemically synthesized polyacryl and polymethacryl amides have lower molecular weight (Simonescu et al. 1983). Acrylonitrile, styrene, ε-caprolactam, and isoprene as well as aryl- and methacryl amides have special optimal times of the polymerization on grinding (Oprea and Popa 1980). For the aryl- and methacrylamides, the polymerization proceeds slowly, usually between 24 and 72 h. After that, some acceleration takes place and the process is completed in 96 h (total).

Chain growth is predominant at the beginning of the process, when mainly unreacted monomer presents in the reaction medium, and the synthesized polymers have not reached sufficient size (critical length) to concentrate the mechanical energy. Molecular weight also is increased near the maximum of the conversion, when most monomer is consumed during the acceleration period. When this maximum is reached, degradation takes place and results in a decrease of molecular weight to a limiting value of 10^3 to 10^4. Hence, the fixed and even reduced molecular weight of polymers is the specific feature of such polymerization.

Tribochemical reactions can be used for creation of thermally stable polymer layers. Such layers form on various surface factors. They stretch in the direction of friction. They are very dense, although amorphous (Simonescu et al. 1983). The layers are more thermally and frictionally stable than the thermally stable polymers obtained by conventional methods (Krasnov et al. 2002). The polymerization discussed can be of interest if amorphous polymers with moderate molecular weights are needed.

Mixtures of methylmethacrylate with styrene could be polymerized mechanochemically by the active species resulting from grinding of quartz (Hasegawa et al. 2002). Interestingly, in this case no electron exoemission was observed. At the initial grinding time, an induction period (about 6 h) took place in which absolutely no polymerization proceeded. The quartz surface produced by grinding participated in the polymerization. The polymerization between different monomer molecules proceeded regularly and alternately. Accordingly, the copolymers formed in such a reaction system were alternating copolymers.

The composition curve of the process is practically the same as the curve obtained when the radical copolymerization of methylmethacrylate and styrene was carried out with azobis(isobutyronitrile) as a radical initiator. The copolymerization in the ground quartz system is undoubtedly radical polymerization. The generation of the initiating radicals is caused by quartz grinding, that is, the scission of the silicon–oxygen–silicon fragments of this inorganic component. Such a conclusion is corroborated by the fact that the copolymerization degree is closely connected to the total surface area of the ground quartz.

When polymerization proceeds in the presence of modifiers, the mechanochemical process enhances cross-linking and, correspondingly, improves the physicochemical properties of the final plastics. For example, mechanochemical treatment of ABS plastic in the presence of toluene diisocyanate improves the thermal oxidative stability of the plastic (Chetverikov et al. 2002).

So, it is obvious that mechanical activation of monomers brings about two competitive processes: the growth of polymer chain and the chain destruction. A number of studies were undertaken to find optimal conditions regarding duration of the polymerization reaction, its temperature, its allowed loading, and so on. An example was presented in the article by Mit' and coauthors (2003).

Krasnov and coworkers (2003) gave a principal example of the combination of tribochemical and conventional methods for preparation of poly(peryleneimides) with improved properties. Polymers of this type are used in the manufacture of films for microelectronics. Conventionally, synthesis of such polymers is conducted in the chlorophenol-phenol mixture at a temperature exceeding 210°C for 12 h. The solid-state polymerization of 4,4'-diaminodiphenyl oxide and piperylene-3,4,9,10-tetracarboxylic dianhydride was mechanically initiated. To prevent destructive processes on prolonged grinding, the mechanochemical action was stopped after 40 min; after that, the polymer obtained was dissolved in phenol and heated to 160°C for 16 h. The resulting polyimide showed perfect characteristics and formed transparent bright red films. It was also demonstrated (Mit' et al. 2004) that the reaction between diaminodiphenyl oxide and pyromellite dianhydride led to oligomeric amido acids with a molecular weight up to 2000 if the synthesis was carried out in a shear-type mixer using ethanol as a dispersing medium.

4.2.2 MECHANOLYSIS AND DEPOLYMERIZATION

The term *mechanolysis* denotes bond scission under mechanical activation. Such a reaction is reversible in principle. As this takes place, the free radicals formed enter the usual free-radical reactions: recombination, decomposition, addition, and substitution. However, the mechanically induced homolysis has some features. First, the implemented energy stretches the polymer backbone chain, which then is cleaved: R–R [R· ⋯ R·] R· + R·. The radicals R· are removed from each other and leave the unit volume. This leaving obeys diffusion law. The rupture of the junction bond is the limiting step of the reaction.

Mechanical stress enhances the mobility of the reacting species. The formation of free space and creation of a favorable arrangement of radicals are crucial for radical recombination. The important specificity of mechanochemical (mostly solid-phase) reactions consists of generation of active species and ensuring mass transfer (to let reagents meet with each other). Two mechanisms of mechanochemical reactions are most likely. First, under the action of mechanical stress, intermixing occurs at a molecular level. Second, the product forms on the surface of macroscopic reacting species.

Formed in the solid phase, the radicals generated recombine so that mechanolysis proceeds as a reversible reaction. However, the term *reversibility* should be applied only to the bond formation between radicals. Namely, the structure of the "recombined" product can be and is different from that of the starting material. Depolymerization of some natural polymers is an example.

Milling of the cellulose carboxymethyl derivatives, chitin and chitosan, at ambient temperature leads to scission of the main polymeric chain. Cleavage of 1,4-glucosidic bonds takes place. In both cases, radicaloid products are formed and recombined. The homolysis and recombination were monitored by electron spin resonance (ESR) (Sasai et al. 2004). Only midsize polymeric chains are formed on radical recombination. Some balance is established between the homolytic depolymerization and the size-limited recombination of the radicals primarily formed.

At this point, it is useful to compare mechanochemical modification of polystyrene, poly(methyl methacrylate), and styrene–acrylonitrile copolymer in a vibratory mill and in an ultrasonic reactor (Boehme et al. 2003). A vibratory mill allows obtaining low molar mass polymers in a short time. However, the advantages of the mill are restricted because of the degradation of the polymers. To generate macroradicals, the ultrasonic reactor is preferred.

With all the vinyl polymers, the active terminal radicals pass into the more stable radicals with the free valence in the middle of the chain (chain radicals). The chain radicals lead to the degradation of the chain by forming C=C bonds:

$$—CH_2—CH_2{}^{\cdot}—CH—CH_2—CH_2— \rightarrow —CH_2—CH{=}CH + {}^{\cdot}CH_2—CH_2—$$

$$\underset{\displaystyle R}{|} \qquad\qquad\qquad \underset{\displaystyle R}{|}$$

According to the general rule of alkene chemistry, the unsaturated bonds are usually formed close to branches. On formation, such radicals react with neighboring macromolecules, abstracting a hydrogen atom from them. Such abstraction begins the chain degradation of polymers. Nevertheless, the number of macromolecules with double bonds is in fact 10^3 times higher than the number of free radicals in the less-stressed regions of the polymer after breaking of the chain reaction (Heinicke 1984).

For polymers, the relative rates of destruction are defined first by the rigidity of their structure. The rigidity increases on passing from flexible-chain polyethylene to more rigid polystyrene and then to polypeptides. Polypeptides possess special rigidity owing to the presence of peptide bonds in the backbone and a dense network of hydrogen bonds. The maximum destruction rates were detected for three-dimensional polymers such as glyceromaleate (Dubinskaya 1999). Because of the nonuniformity of the stress distribution inside the polymer, the overloaded bonds are first broken after the application of stress.

A treatment device introduces special effects in mechanolysis. For instance, it is impossible to obtain a cross-linked polyethylene with maleic anhydride in the usual extruders. In this case, there are too few macroradicals available for reaction because of insufficient polymer degradation. In the disk-type extruder, a higher stress gradient is achievable, more macroradicals are generated, and intensive cross-linking between highly chlorinated polyethylene and maleic anhydride or methyl methacrylate can be obtained (Heinicke 1984; Zhao et al. 2002, 2003).

After formation, the macroradicals not only can enter recombination or disproportionation reactions, but also can be involved in secondary polymeric transformation leading to block copolymers and cross-linked or grafted polymers. However, the products formed are again converted during mechanolysis. Thus, for instance, three-dimensional structures degrade to a mixture of grafted, cross-linked, and block polymers (Oprea and Simonescu 1972).

During mechanolysis of poly(methylmethacrylate), the radical concentration gradually increased to a maximum and then gradually decreased (Kondo et al. 2004). It was suggested that the radical disproportionation or recombination progressed together with radical formation. Regarding molecular mass of the starting polymer, it decreased exponentially toward the limiting value, and the limiting mass was larger with decreases in mechanical energy. All of these data on poly(methylmethacrylate) are in accordance with the regularities expressed in this and preceding sections.

Carboxylated polymers can be prepared by mechanical treatment of frozen polymer solutions in acrylic acid (Heinicke 1984). The reaction mechanism is based on the initiation of polymerization of the frozen monomer by free macroradicals formed during mechanolysis of the starting polymer. Depending on the type of polymer, mixed, grafted, and block polymers with a linear or spatial structure are obtained. What is important is that the solid-phase reaction runs with a relatively high rate. For instance, in the polyamide reactive system with acrylic acid, the tribochemical reaction leading to the copolymer is completed after a 60-sec treatment time. As a rule, the mechanical activation of polymers is mainly carried out in a dry state because the structural imperfections appear most likely here.

Tribochemistry also was applied as an economical method of rubber devulcanization. Devulcanization is the process of rupture, entirely or partially, of carbon–carbon, carbon–sulfur, or sulfur–sulfur bonds in the chemical network formed during polymer cross-linking in the process of rubber manufacturing. Each year, about 2 billion tires are discarded throughout the world, which represents a major loss of an important resource. Known methods of devulcanization involve high energy consumption, complicated equipment, and toxic reagents and produce toxic by-products or lead to devulcanized compounds that are not industrially acceptable. As a result, the cost of known methods of devulcanization is too high.

Mechanical grinding of rubber in the presence of a chemical additive brings about devulcanization at a very low cost (Sangari et al. 2003). During mechanochemical treatment, the direct breakage of the carbon–carbon backbone chain takes place alongside the breakage of carbon sulfur and sulfur–sulfur bonds. Free radicals form and recombine. Chemical additives are used to control the recombination. This provides compounds that can be molded and revulcanized within the conventional rubber manufacturing process. The revulcanized samples showed good mechanical properties for further industrial applications.

4.3 REPRESENTATIVE EXAMPLES OF MECHANICALLY INDUCED ORGANIC REACTIONS

There are numerous chemical systems that show a different course of reaction during the mechanical stress in relation to thermal conditions. This section provides examples.

4.3.1 THE NEWBORN SURFACE OF DULL METALS
IN ORGANIC SYNTHESIS

4.3.1.1 Bismuth with Nitroarenes

Bismuth is a typical example of a dull metal. At normal temperatures, it does not react with water or oxygen. It finds rather restrictive application to organic synthesis, including reduction reactions. Nevertheless, the metal is cheap and not toxic; some bismuth salts are orally taken as medicines for intestinal disorders (Briand and Burford 1999). Grinding is attractive for activating this metal for reactions with organic substrates.

When the metal is mechanically crushed, the newborn metal surface is highly activated. Because its ionization potential is high enough, the surface does not immediately react with atmospheric oxygen and moisture to form an undesirable film of metal oxide/hydroxide. The activated metal surface survives for a long enough to react with neighboring molecular species other than those of atmospheric origin.

Here, we consider nitroarenes the species neighboring the activated bismuth. A nitroarene reactant, bismuth shots, and a trace amount of hexane were shaken with stainless balls in a stainless cylinder to form a diarylazoxy compound in an almost-quantitative yield (Wada et al. 2002). According to the authors, a nitroarene was adsorbed and deoxygenated on the newborn bismuth surface to form a nitrosoarene as an initial product. The two nitrosoarenes gave a diarylazoxy derivative.

The methodology described has been successfully extended to the single-step synthesis of long-chain 4-alkoxyazoxyarenes from the corresponding nitro compounds (Wada et al. 2002). The reaction was clean, and the yield based on conversion was almost quantitative. These molecules of vast elongation are key materials for electronic devices based on their liquid crystalline properties. Traditional methods of their synthesis (by wet reactions) lead to product mixtures that contain compounds that differ in a degree of the nitro group reduction. It is difficult to separate the azoxy product from those mixtures. True, the alkyl chain length does not seem to have much influence on the product yield in ordinary solution chemistry. The solvent-free mechanochemical reaction is essentially a surface event. The size effect is understandably important in adsorption phenomena. In the case under consideration, the unsatisfactory conversion was observed with nitroarenes with an alkoxy chain longer than eight carbon units. This restriction may be attributed to inefficient mixing, but it may also reflect some role played by the longer alkoxy chain in placing a long-chain nitrobenzene on the structured solid surface. Despite the restriction mentioned, this bismuth-mediated procedure is a valuable expansion of azoxy liquid crystal technology.

Tris(2-nitrophenyl)bismuthane can be prepared by grinding 2-nitroiodobenzene with bismuth (to acquire a fresh surface). To be activated, the reaction requires the presence of metallic copper and cuprous iodide (Urano et al. 2003). This transformation is common for iodobenzene derivatives bearing electron-withdrawing groups at the ortho position. For instance, the fluoro, bromo, or chloro derivatives at the ortho position give rise to the corresponding triarylbismuthanes in greater than 80% yields. The authors suggested that the reaction proceeded through formation of an aryl copper species, which underwent ligand exchange with a bismuth atom on the fresh metal surface. Triarylbismuthane formed via either stepwise arylation or disproportionation of some arylbismuth species. The overall reaction was as follows:

$$[3(o–RC_6H_4I) + Bi\ (Cu, CuI, CaCO_3)]\ (ball\ milling) \rightarrow (o–RC_6H_4)_3Bi$$

All the bismuthanes are difficult to access by conventional wet routes.

4.3.1.2 Tin with Benzyl Halides

Under exclusively thermal initiation conditions, benzyl halides react with tin to give poly(phenylmethylene). Compared to this, with mechanical stress dibenzyl dichlorostannane is formed (Grohn and Paudert 1963). Namely, tin and benzyl chloride are fed together with porcelain balls into the milling vessel. After exchanging the air atmosphere for nitrogen, the reaction mixture is stressed tribochemically in a vibration mill for 3 h. The final product (dibenzyl dichlorostannane) is formed almost quantitatively. The following transformations take place:

$$Sn + 2PhCH_2Cl \rightarrow SnCl_2 + 2PhCH_2^{\cdot}$$

$$Sn + 2PhCH_2^{\cdot} \rightarrow (PhCH_2)_2Sn$$

$$(PhCH_2)_2Sn + 2PhCH_2Cl \rightarrow (PhCH_2)_2SnCl_2 + 2PhCH_2^{\cdot}$$

$$2PhCH_2^{\cdot} + SnCl_2 \rightarrow (PhCH_2)_2SnCl_2$$

4.3.1.3 Aluminum/Hydrogen Plus Olefins

Reactions of olefins with aluminum in hydrogen atmosphere result in formation of aluminum alkyls. The reactions are heterophaseous. They are complicated by mass transfer, oxidation of the metal surface, and the overall dull nature of the metal. All these complications are easily circumvented if mechanochemical activation is used. Mechanochemical activation of aluminum by the addition of nickel or titanium (5–10 or 1%, respectively) and sodium chloride as a supporting agent was studied by Lukashevich et al. (2002). At less than 10 MPa hydrogen, grinding of 1-alkenes (1-heptene, styrene, or dihydromyrcene) with aluminum at 120°C led to organoaluminum derivatives. *In situ* oxidized with air and hydrolyzed, these derivatives gave 1-alkanols (1-heptanol, 2-cyclohexyl ethanol, or citronellol with yields of 60, 21, or 65%, respectively).

Citronellol is a scent additive for soups and cosmetics; it is expensive at present. Lukashevich and coauthors (2002) and Goiidin and colleagues (2003) offered an economy-type industrial technology for citronellol manufacturing based on turpentine components (Scheme 4.1).

4.3.2 Reactions of Triphenylphosphine with Organic Bromides

Based on the advantages of mechanochemical synthesis, preparation of 1-(4-bromophenyl)-2-(2-naphthyl)ethylene was patented (Pecharsky et al. 2003). Milling is conducted in the presence of potash in a controlled atmosphere without any solvent

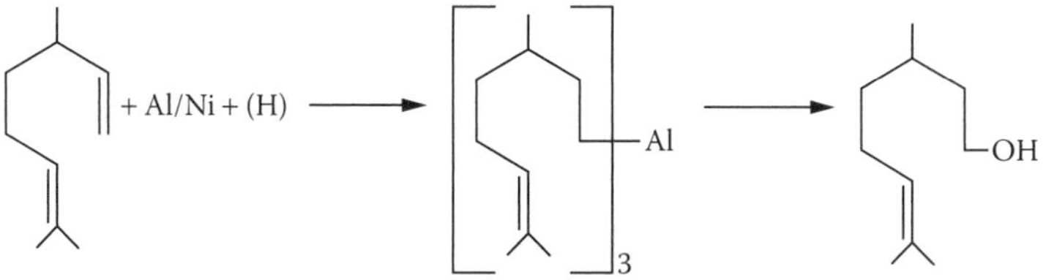

SCHEME 4.1

for 8 h. Reactants and the product (92% yield) are depicted in Scheme 4.2. The reaction of Scheme 4.2 is Wittig condensation, which proceeds through a phosphorus ylide.

Phosphonium salts have been prepared during high-energy ball milling of triphenylphosphine with solid organic bromides (Balema et al. 2002a). The reactions occur at ambient conditions with no solvent. The 2-bromo-2-phenylacetophenone case is typical: The reaction in a solution usually produces a mixture containing both the C-phosphorylated and O-phosphorylated compounds (Borowitz et al. 1969). The solvent-free ball milling induces regioselective transformation. In the mechanical process (Balema et al. 2002a), only the thermodynamically favorable C-phosphorylated product forms (the yield is 99%):

$$Ph_3P + BrCH(Ph)C(O)Ph \rightarrow [Ph_3P^+CH(Ph)C(O)Ph]Br^-$$

One of the probable reaction mechanisms assumes that when low-melting halides react with triphenylphosphine, low-melting eutectics are formed during ball milling. In this case, the reactions may occur in the melt. This melt forms locally and momentarily in the areas where the rapidly moving balls collide with both the walls of the reaction vial and one other. According to Balema et al. (2002c), the local temperature in a material during ball milling does not exceed 110°C.

The material nature of the vial and balls can be crucial for the reaction. Thus, (2-naphthyl)methylene triphenylphosphonium bromide [$Ph_3P^+CH_2$(2-Naph)Br$^-$] easily forms during ball milling of triphenylphosphine and 2-bromomethylnaphthalene in hardened-steel equipment for 1 h (the yield is 95%). The transformation of triphenylphosphine and 1,3-dibromopropane into propane-1,3-diyl bis(triphenylphosphonium)

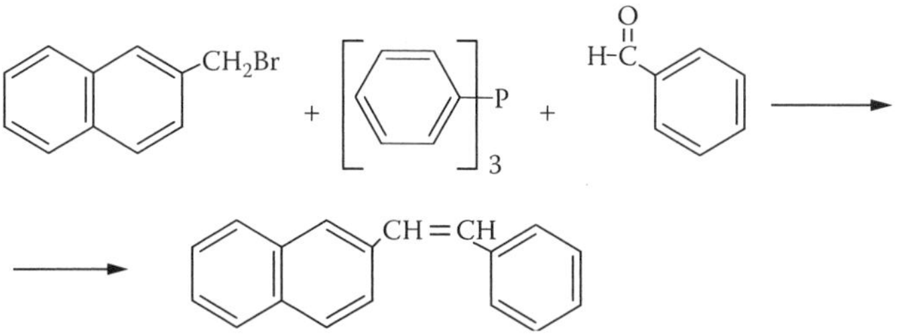

SCHEME 4.2

dibromide [(Ph$_3$P$^+$CH$_2$CH$_2$CH$_2$P$^+$Ph$_3$)·2Br$^-$] proceeds at stronger conditions, namely, by mechanical processing in a tungsten carbide vial with tungsten carbide balls for 12.5 h (the yield is 51%). In comparison with steel balls, heavier tungsten carbide balls increase the input of mechanical energy into the system (Suryanarayana 2001).

4.3.3 REACTIONS OF ORGANYLARSONIUM OR DICHOROIODATE(I) WITH OLEFINS AND AROMATICS

Ren, Cao, Ding, and Shi (2004) described a novel route for highly stereoselective synthesis of *cis*-1-carbomethoxy-2-aryl-3,3-dicyanocyclopropane by grinding. A mixture of methoxycarbonylmethyl triphenylarsonium bromide, arylidenemalononitrile, potassium carbonate, and several drops of water was ground at room temperature in a glass mortar with a glass pestle for 30 min. After column chromatography, triphenyl arsine was recovered, and products of cyclopropanation were obtained at 70–90% yields. The process is stereoselective, simple, efficient, and environmentally benign. Compared to the same reaction in dimetoxyethane, the solvent-free process proceeds 12 times faster and leads to higher product yields.

The solid-state reaction of *trans*-stilbene with potassium dichoroiodate(I) in a vibrating steel ball mill gives 1-iodo-2-chloro-1,2-diphenylethane at an 85% yield. Scheme 4.3 depicts this reaction as a dry milling process. Such a reaction absolutely does not occur during the day stirring of the reactants in the carbon tetrachloride solution (Sereda et al. 1996). Mechanical destruction of the solid and brittle crystals of *trans*-stilbene results in emission of local energy, which initiates the iodochlorination.

In contrast to *trans*-stilbene, methyl cinnamate underwent iodochlorination by potassium dichoroiodate(I) in carbon tetrachloride (a wet reaction in Scheme 4.3), but the solid-state reaction on vibromilling did not take place (Sereda et al. 1996). During mechanical treatment of the short-brittle and low-melting crystals of methyl cinnamate, the mechanical energy was consumed for their melting, not to activate the iodochlorination.

The mechanical properties of the starting olefin exert the decisive influence on the solid-phase process. Scheme 4.3 juxtaposes both reactions considered.

Hajipour and colleagues (2002) described the iodination capability of tetramethylammonium dichloroiodate(I) regarding aromatic compounds. The reaction is

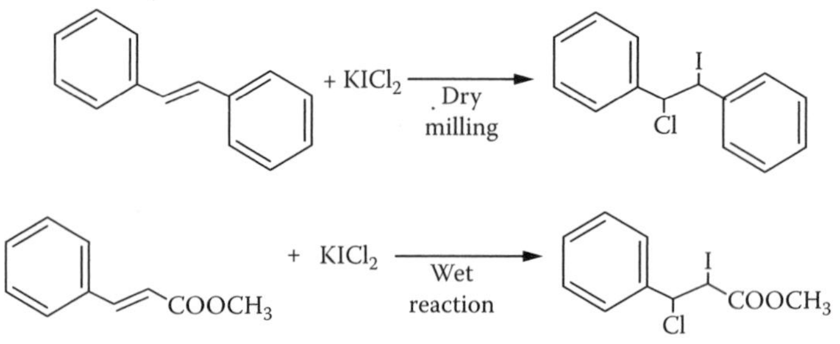

SCHEME 4.3

initiated with mortar-and-pestle grinding of the starting compounds and leads, after keeping the mixture at 5–20 min at room temperature, to iodoaromatics. Scaled-up experiments showed that the yields are excellent. The process is solvent free. This is also a case of environmentally friendly iodination.

4.3.4 REACTIONS OF METAL FLUORIDES WITH POLYCHLOROAROMATICS

Fluoroaromatics are very important compounds for many applications. Their manufacturing requires severe conditions, with excessive consumption of inorganic reactants at high pressures and temperatures as well as long duration (60 h and more). When solvents (formamide, sulfolane) are used, cost and environmental problems emerge. The solid-phase synthesis of polyfluorinated derivatives of benzene, pyridine, and naphthalene was accomplished by treating the polychloro derivatives with potassium fluoride or potassium calcium trifluoride in a planetary-centrifugal mill (Dushkin et al. 2001). With potassium fluoride, fluorination of pentachloro pyridine needs 2 h to reach 40% yield; with potassium calcium trifluoride (derived from the mixture of potassium and calcium fluorides), the reaction is completed in only 0.5 h. Using potassium fluoride and octachloro naphthalene, Dushkin et al. (2001) studied the temperature effect on the mechanochemical reaction. The transformation increased when the water temperature in the external heat exchanger of the mill increased from 10 to 25°C. However, further increasing the temperature retarded the reaction. The temperature increase led to an increase in plasticity of the organic reactant that lowered the mechanical stress and the activation degree.

4.3.5 NEUTRALIZATION AND ESTERIFICATION

The mechanochemical variant of neutralization provides an opportunity to consider effects of the neutralizing agent. Zaitsev et al. (2001) compared two systems: mechanochemical neutralization of acetyl salicylic acid with sodium carbonate and mechanochemical activation of a mixture of the same acid with calcium carbonate. In the first system, sodium salicylate formed after 12 h of activation. No calcium salicylate was formed in the second system. Sodium salicylate from the first system had enhanced solubility in water. The solubility was much higher than that of the sodium salt obtained after the conventional neutralization of salicylic acid with sodium hydroxide in aqueous solutions. The mechanochemically prepared analog was patented as aspinate. Aspinate is advantageous in comparison with other pain relief medications (see Zaitsev et al. 2001). Boldyrev (1996) noted some industrial merits of this mechanochemical process. The traditional scheme includes six technological stages and requires 70 h. One needs 500 l water and 100 l ethanol to produce 500 kg salicylate. The same amount can be produced under mechanochemical conditions in one stage after 7 h from the solid starting materials without any solvent.

A similar example is provided by the synthesis of sodium benzoate, another important product of the pharmaceutical industry. Traditionally, it is produced by the neutralization of benzoic acid by soda in aqueous solution. A standard technological cycle consists of six stages. Production of 500 kg benzoate requires 3000 l water. A standard duration of the cycle is 60 h. The same amount of sodium benzoate can be produced by mechanical treatment of the mixture of solid powders of benzoic acid and soda for only 5–8 h (Boldyrev 1996). The consumption (and the contamination) of enormous amounts of water is excluded. The acceleration of the technological rate of the process is also obvious.

Du and associates (2002) prepared calcium ascorbate by mechanical synthesis from ascorbic acid and active calcium. The product is used in animal nutrition. Compared with other methods of manufacture, the costs of production and equipment investment were very low. The reaction duration was short, and the quality of the product was excellent. This technology is suitable for industrial application.

Ionic fluorides are widely used as basic reagents in organic chemistry, and fluoride ion affinities provide a novel scale for Lewis acidity (Christe et al. 2000). With H-acids, the fluoride anion is capable of detaching a proton, forming the very stable hydrogen difluoride anion HF_2^- (Pimentel 1951). In the solid state, even weak acid such as nicotinic acid ($k_a = 1.4 \cdot 10^{-5}$) reacts mechanochemically with potassium fluoride to form potassium nicotinate and KHF_2 (Fernandez-Bertran and Reguera 1998; Fernandez-Bertran et al. 2002):

$$RCOOH + 2KF \rightarrow RCOOK + KHF_2$$

With the carboxylic groups of hemin, potassium fluoride mechanochemically reacts according to the same scheme. However, lithium and sodium fluorides are inert when milled with hemin (Paneque et al. 2002). Lithium fluoride exists as a strong ion pair, which explains its inertness in mechanochemical reactions. Lithium fluoride is inert even with the very active oxalic acid (Fernadez-Bertran and Reguera 1998). On other hand, the carboxylic groups of hemin are not acidic enough to participate in a proton transfer to sodium fluoride: When sodium fluoride is milled with oxalic acid, monosodium oxalate does form (Paneque et al. 2002). Consequently, sodium fluoride can remove only very acidic protons.

Tetraaryl borate salts with bulky organic cations are used as components of high-performance catalytic systems for industrial manufacture of polyolefins. Borisov and others (2004) claimed a method for preparing these salts by milling of the potassium tetraphenylborate mixture with the equimolar amount of tris(pentafluorophenyl)methyl bromide in controlled atmosphere for 1 h. The yield of $(C_6H_5)_3CB(C_6F_5)_4$ product (isolated) was 80%. In a comparison example, refluxing the same reactants in hexane for 12 h gave only 48% of $(C_6H_5)_3CB(C_6F_5)_4$.

Cobaltocene dicarboxylic acid represents an unusual case of interaction with alkali halides on manual grinding in an agate mortar. This compound is an ion of $[Co^{III}(-C_5H_4COOH)(-C_5H_4COO^-)]$ structure. In solid state, this ion exists as a trimer bounded with hydrogen bonds between COOH and COO^- fragments. Grinding with cesium iodide leads to the formation of an inclusion complex in which two cesium cations are located in the cavities in the trimer formed (Braga and coauthors 2004). The situation resembles inclusion of an alkali cation in a crown ether cavity.

As in the case of neutralization, the simple reaction of esterification is here considered regarding features of the mechanochemical process. When milled, cellulose and maleated polypropylene interact according to esterification. It is known that hydroxyl groups of cellulose are connected by intramolecular and intermolecular hydrogen bonds. Milling leads to considerable collapse of such bonds. The collapse generates many free hydroxyl groups in the substrate. The resultant OH groups on the cellulose surface are very reactive. For this reason, the esterification proceeds much more deeply than that in the conventional melt-mixing process. The product obtained has stronger exploitation properties. In particular, it acquires improved tensile strength and, importantly, better compatibility with a hydrophobic polypropylene matrix (Qiu et al. 2004).

Sometimes, mechanical induction allows choosing a more convenient manner of esterification and overcome such obstacles as with too high viscosity of a reaction medium. Thus, conventional manufacturing of pentaerythritol phosphate alcohol consists of the reaction between pentaerythritol and phosphorus oxychloride in dioxane. (The alcohol formed is further used to produce flame retardants and plasticizers.) Hydrogen chloride is generated in this method, and a large amount of water is required to wash away HCl from the product mixture. Moreover, an excessive amount of $POCl_3$ is consumed in this method, which results in a residue solution containing unused and highly reactive phosphorus oxychloride. Ma and coworkers (2004) claimed a method for preparing pentaerythritol phosphate alcohol; the method involved ball milling a mixture of phosphorus pentoxide, pentaerythritol, and toluene as a solvent in the presence of magnesium chloride at 90–150°C. Although not dissolved in toluene, pentaerythritol turns into a molten state when the temperature of the mixture rises to 90°C and can react with phosphorus pentoxide in the suspension. Because the reaction proceeds in suspension, strong agitation is needed. Because pentaerythritol in its molten state has extremely high viscosity, it is difficult to agitate the suspension. The keys of the method consist of using a ball mill as a reaction vessel and preliminary heating of the solvent to 90°C. The reaction is carried out for 6 h. The obtained product mixture contains approximately 80% pentaerythritol phosphate alcohol and approximately 20% phosphoric acid. The yield of the alcohol is more than 95%.

4.3.6 ACYLATION OF AMINES

Dry mechanochemical technology, because it is more convenient ecologically, sometimes allows obtaining the desired product with a higher yield and a higher rate. For example, the standard method of manufacturing phthalazole in the pharmaceutical industry is to heat aqueous and alcoholic solutions of sulfathiazole and phthalic anhydride in the presence of acid catalysts or to melt both reactants. The product is unavoidably contaminated by phthalimide and phthaloyl bisamide in the case of melting. For reaction in solutions, additional contaminations are diethyl esters of phthalic acid. Unlike these methods, milling a mixture of the reactants allows obtaining rather pure phthalazole free from contaminations. Benzoic acid accelerates the reaction (Chuev et al. 1989) (see Scheme 4.4).

Detailed studies of the mechanism of this acylation were performed by Mikhailenko and coauthors (2004a, 2004b). In the authors' opinion, local evolution of heat at the contacts causes sublimation of phthalic anhydride onto the surface of sulfathiazole crystals. Grinding permits continuous renewal of the sulfathiazole

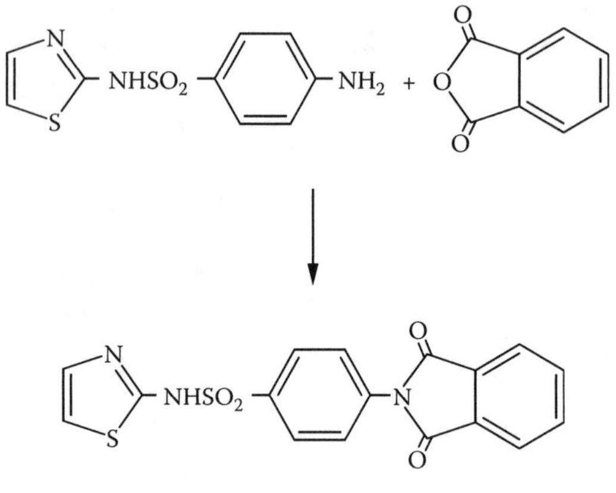

SCHEME 4.4

crystal surface and permanent removal of phthalazole formed from the reaction region, providing a fresh opportunity for the reaction to proceed. The acceleration effect of benzoic acid was explained by changing of the rheological properties of the mixture, which helps to grind sulfathiazole particles.

4.3.7 DEHALOGENATION OF PARENT ORGANIC COMPOUNDS

Dehalogenation of parent organic compound reactions are important from the standpoint of environmentally related problems. In particular, chloro-organic compounds have been linked to increased risk of several types of cancer. Despite reductions in their use, they remain one of the most important groups of persistent pollutants to which humans are exposed, primarily through dietary intake. Many works have described mechanochemical degradation of halogenated (mostly chlorinated) organic compounds by high-energy ball milling. Calcium and magnesium powders and their oxides were used as reagents for organohalides (Rowlands et al. 1994).

Trichlorobenzene, as an example of a chlorinated compound, was decomposed by dry grinding with calcium oxide. Calcium chloride and carbon were the main final products, achieving the goal of transforming toxic organics into inorganics that are safe and can be stored without problem or can be used in appropriate fields (Tanaka, Zhang, Mizukami, and Saito 2003; Tanaka, Zhang, and Saito 2003a,b).

Evidence was presented that mechanochemical destruction of pesticide DDT by steel ball milling in the presence of calcium oxide eventually leads to graphite (Hall et al. 1996). The method is applicable even to treatment of soils contaminated with DDT and polychorobenzenes (Masuda and Masame 2001a). The first step of the mechanochemical reaction is dechlorination; aromatic ring cleavage and polymerization take place next, and complete destruction of organic intermediates finalizes the treatment (Nomura et al. 2002).

Calcium oxide catches chlorine removed from the parent chlorinated compound and forms calcium chloride. In addition, calcium oxide oxidizes the organic compounds. If poly(chloroarylamide) (DuPont's Aramid) is ground with an excess amount of calcium oxide, the main products are amorphous carbon, calcium chloride, and calcium nitrate. The starting material is decomposed quantitatively (Tanaka, Zhang, and Saito 2003a,b; Tanaka et al. 2004). The intensity of the mechanical treatment defines the decomposition kinetics. To reach a high degree of transformation in vibration mills, 6–18 h are needed, whereas only several minutes are sufficient if planetary centrifugal mills are used (Korolev et al. 2003).

Calcium oxide (quicklime) is, of course, not the most convenient reagent. Therefore, alumina or silica treatment was proposed for neutralization of soils contaminated with chlorinated dibenzodioxin or dibenzofuran (Masuda and Masame 2001b). The mechanochemical contact of polychorobiphenyls and birnesite (δ-MnO$_2$) removes the pollutant. The dechlorination degree depends on the number of chlorine atoms and their positions in biphenyl rings. The more chlorines there are, the longer the milling time that is needed. Regarding mutual disposition of chlorines, ortho dichlorobiphenyls are more reactive than the meta or para isomers (Pizzigallo et al. 2004a).

Birnesite activity at milling can be enforced by adding humic acids, as has been shown for pentachlorophenol dechlorination (Pizzigallo et al. 2004b). The authors noted: "Humic acids contain indigenous free radicals that may be involved to a various extent in chemical processes. However this field of research is open and warrants further study."

Milling of 1,2,3,4-tetrachlorodibenzodioxin in hydrogen atmosphere (1–1.5 MPa) with Mg$_2$FeH$_6$ catalyst leads to tetrachloro- and pentachlorobenzenes. The reaction products contain no dioxin (at a determination accuracy up to 0.0001%). According to the authors (Molchanov, Goiidin, et al. 2002), this method is the most effective and cheapest among all known procedures for dioxin deactivation.

Dehalogenation of bromo- (Saito et al. 2002) or fluoropolymers (Nagata et al. 2001) without heat treatment for recycling wastes was developed. Mechanochemical treatment of pulverized halopolymers with alkali metal hydroxides proceeds at a normal temperature. The resulting products are recyclable as fuels. In the case of sodium fluoride, it is recycled as a substitute for fluorite (calcium fluoride) in the inorganic fluorine industry.

4.3.8 COMPLEXATION OF ORGANIC LIGANDS TO METALS

When mechanical stress is exerted on coordination compounds in a solid state, their constituent ligands are subjected to distortion. This causes a change in the strength and anisotropy of the ligand field. In turn, it leads to a considerable modification of reactivity. As one simple example, the synthesis of [Fe^{II}(phen)$_3$]Cl$_2$·nH$_2$O should be mentioned. Preliminary milling of FeCl$_2$·4H$_2$O activates the aqua complex so that it readily reacts with 1,10-phenanthroline (phen). After 3 min of subsequent milling, the yield of the organometallic complex is quantitative. Without preliminary activation of FeCl$_2$·4H$_2$O, milling for 3 h is needed to complete the reaction. On the other hand, preliminary milling of phenathroline alone did not bring about any change in the subsequent mechanochemical reaction with FeCl$_2$·4H$_2$O (Senna 2002). In contrast to conventional synthetic methods, the mechanochemical free-of-solvent procedure is quite easy and proceeds in mild conditions. This is truly green chemistry.

Other examples of such mechanochemical reactions are solid-state preparation of praseodymium acetyl acetonates (Zaitseva et al. 1998), europium quinaldinate, phthalate, or cinnamate (Kalinovskaya and Karasev 1998, 2003) and transition metal complexes with tris(pyrazolyl)borate (Kolotilov et al. 2004). As pointed out, an induction period is required for the reaction to begin. The more amounts of reactants are introduced in production, the longer the induction period is. The grinding-initiated reaction between dimethylglyoxime and cupric acetate is an example (Hihara et al. 2004).

Mechanical activation leads to autothermal initiation of complexation. Such exothermal effect has been definitely revealed by differential thermal analysis. Evidently, the effect originates a self-propagation reaction sequence (Makhaev et al. 1998; Petrova et al. 2001). The final complexes are formed on continuation of the mechanical action (Petrova et al. 2004). Loss of crystallinity during milling probably also plays a role. To accumulate such effects, some time is needed.

Ball milling of platinum dichloride and triphenyl phosphine for 1 h leads to *cis*-bis(triphenylphosphine)platinum(II) dichloride at a 98% yield (Balema et al. 2002b):

$$PtCl_2 + (C_6H_5)_3P \rightarrow cis\text{-}[(C_6H_5)_3P]_2PtCl_2$$

The mechanical processing of *cis*-bis(triphenylphosphine)platinum(II) dichloride with anhydrous potassium carbonate produces the carbonate complex at a 70% yield (Balema et al. 2002b):

$$cis\text{-}[(C_6H_5)_3P]_2PtCl_2 + K_2CO_3 \rightarrow cis\text{-}[(C_6H_5)_3P]_2PtCO_3 + 2KCl$$

In the absence of mechanical treatment, this reaction does not proceed. One can propose that the reactions may occur in the melt. The melt possibly forms locally and momentarily in areas where balls collide with the walls of the reaction vial and with each other. Such a case is considered in Section 4.3.2. Remember, Balema and coauthors (2002c) showed that the local temperature in a material during mechanical processing (with experimental conditions identical to those used in the current study) does not exceed 110°C. Thus, it is unlikely that the complex $[(C_6H_5)_3P]_2PtCO_3$ forms as a result of a liquid-phase reaction between the chloride complex (193°C mp) and anhydrous potassium carbonate (891°C mp). On other hand, the possibility of the reaction between the transient triphenyl phosphine melt (79–82°C mp) and solid platinum dichloride (581°C mp) could not be entirely excluded. Such a reaction had been described, which led to the 4:1 mixture of the *cis* and *trans* isomers of bis(triphenylphosphine)platinum(II) dichloride (Gillard and Pilbrow 1974). However, Balema et al. (2002b) obtained the cis isomer exclusively. Consequently, occurrence of the mechanism with transient melting is doubtful.

Truly solid-phase mechanochemical transformation represents another possible mechanism. It was experimentally established (Balema et al. 2002b) that the reactants lose crystallinity and become essentially amorphous powders during mechanical processing. So, ball milling enables interactions of reacting centers in the solid state by first breaking the crystallinity of the reactants and then providing mass transfer in the absence of a solvent.

The mechanically activated solid-phase reaction is the sole way to prepare metal diketonates free from solvent molecules coordinated to the central metal. On milling, zinc dichloride and sodium hexafluoroacetylacetonate give rise to the formation of zinc bis(hexafluoroacetylacetonate) at a 70% yield (Petrova et al. 2002). There are examples of complexation caused by short-term grinding only. For instance, grinding of $Ni(NO_3)_2 \cdot 6H_2O$ with 1,10-phenathroline results in the formation of the nickel–phenanthroline complex in 2 min. Grinding with sodium calix[4]arene sulfonate leads to the corresponding inclusion compound containing this nickel complex (Nichols et al. 2001).

Sometimes, pestle-and-mortar grinding provokes not only ligand exchange but also a change in the oxidation state of a central metal. Mostafa and Abdel-Rahman (2000) observed such changes during routine preparation of KBr pellets (disks) containing metallocomplexes for recording infrared spectra. Grinding facilitates both complexation and redox reactions. This can cause serious errors in studies using infrared spectroscopy.

The problem of mechanochemical synthesis of polymers with transition metal complexes grafted to linear chains are under scrutiny as candidates for new technologies (see the review by Pomogailo 2000). By elastic wave pulse activation, Aleksandrov and coauthors (2003) synthesized a polymer containing binuclear niobium clusters grafted to linear polyethylene chains (Scheme 4.5).

To synthesize the polymer depicted on Scheme 4.5, niobium clusters were first prepared under the action of elastic wave pulses on a pressurized solid-phase mixture of lithium niobium phosphate powder with 3,6-bis(*tert*-butyl) pyrocatechol and 3,6-bis(*tert*-butyl)-1,2-benzoquinone. The cluster products thus obtained were extracted with toluene. After evaporation of the solvent, the clusters were introduced into a polyethylene matrix under exposure of elastic wave pulses. The clusters were grafted to the hydrogen detachment site of the polymer (see Scheme 4.5). Interestingly, while the clusters are paramagnetic, the grafted polymer is silent in the sense of ESR spectra (Aleksandrov et al. 2003).

It was mechanochemical synthesis that permitted hemin to coordinate arginine or imidazole. Hemin is the central part of hemoglobin and myoglobin. Myoglobin is active in the muscle, where it stores oxygen and releases it when needed. Hemoglobin is contained in red blood cells and facilitates oxygen transport. Both arginine and imidazoles are bases. Hemin is a cyclic organic molecule made of four linked,

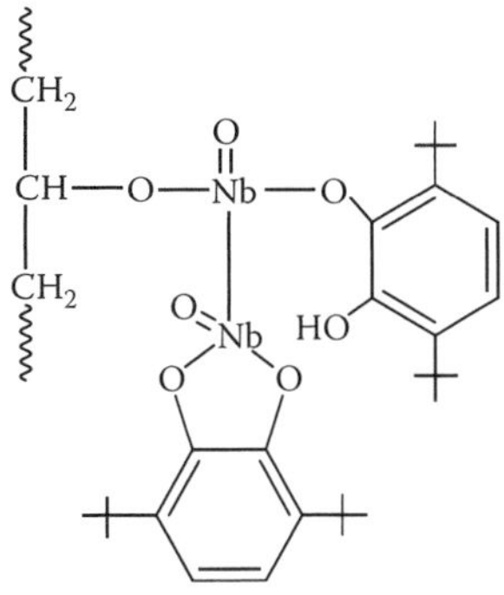

SCHEME 4.5

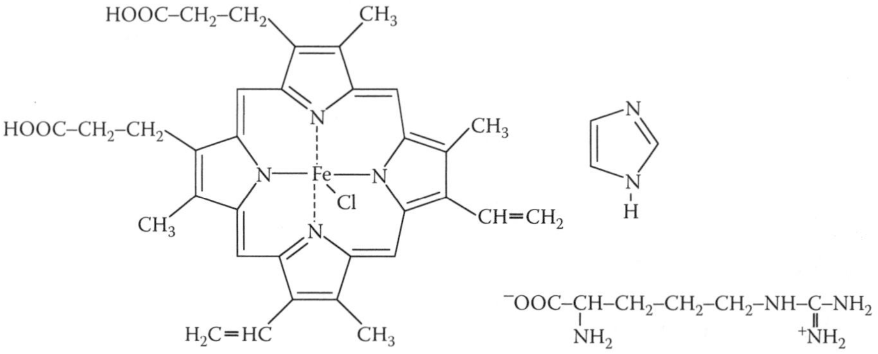

SCHEME 4.6

substituted pyrole units surrounding an iron atom. Scheme 4.6 depicts hemin, imidazole, and arginine.

The central iron atom of hemin is in the five-coordination state. The sixth position remains available for O_2 coordination, the basic step of the respiratory process. Imidazole–hemin complexes are usually studied as simple models of hemoproteins.

The complex between hemin and arginine is an acting source of effective drugs for the treatment of acute porphyria attacks. (When an organism encounters problems in iron absorption, these drugs are administered to correct the iron deficiency.) Methods of preparation of complexes between hemin and arginine or imidazole in solution present problems. Hemin exhibits poor solubility in water or organic solvents and undergoes dimerization even at very low concentrations. In diluted solutions, only complexation at the iron central atom was fixed. At strong dilution, the complexes under consideration dissociate. Other coordination centers of hemin hold the molecular ligands more weakly, and the corresponding coordination bonds, if they exist, are cleaved on dilution. Knowledge of the behavior of all of the reaction centers is significant for some important aspects of hemoglobin functioning. All the obstacles were circumvented when the hemin complexation was performed mechanochemically by manual grinding of hemin with an excess of a ligand (Paneque et al. 2001, 2003). Hemin transforms into the complexes in quantitative yields. The complexes are stable and differ from hemin in their high solubility. Their iron atoms are shielded. This inhibits the formation of iron–iron dimers and coordination between the hemin iron and the carboxylic groups of a neighboring hemin molecule. Coordination of the mentioned types does take place in the case of hemin before its complexation.

According to Moessbauer, infrared, and ESR spectra, one arginine molecule coordinates to the hemin iron atom, and the two others interact with the peripheral acid groups. Regarding imidazole, it forms two different complexes with hemin. The first complex contains two imidazoles bound to iron at axial positions. In the second complex, two imidazoles are bound to iron, and two more are connected with the carboxyl groups in the periphery of hemin. The carboxylic binding is not observed in the corresponding wet reactions.

Finally, one very important technical application of mechanically induced complexation should be considered. Preparation of ceramics containing yttrium, zirconium, and aluminum requires very fine and unimodal mixing. For this purpose, attrition milling of the yttria-stabilized zirconia mixture with aluminum is used. During the operation, both shear and press forces act on the particles by the milling media (namely, tetragonal zirconia polycrystalline milling balls). Because of the malleability of metallic aluminum, its particles could only be deformed and flattened by the milling forces rather than be broken up immediately. At the same time, the hard and smaller yttria-stabilized zirconia particles are compacted with the aluminum flakes. This sets limits to the mixture disintegration.

To circumvent the obstacle, Yuan and coworkers (2004) proposed a suspension process of attrition milling. As a liquid medium for the suspension, they used acetylacetone. This led to homogenization and diminution in the ceramic mixture size. The chemical background of the result deserves to be explained. Acetyl acetone, $CH_3-CO-CH_2-CO-CH_3$, contains the central methylene group surrounded by two carbonyls. This methylene group is acidic. Complexation of acetylacetone to metals leads to elimination of a proton. For the considered cases, the equations are as follows:

$$4Al + 12(CH_3CO-CH_2-COCH_3) + 3O_2 \rightarrow 6H^+ + 4Al^{3+}(CH_3CO-CH^--COCH_3)_3 + 6OH^-$$

$$ZrO_2 + 4(CH_3CO-CH_2-COCH_3) \rightarrow 2H^+ + Zr^{4+}(CH_3CO-CH^--COCH_3)_4 + 2OH^-$$

The main point of these equations consists of releasing large amount of protons. The free protons are absorbed at the particle surface, and subsequently yttria-stabilized zirconia and aluminum participants are charged positively. Electrostatic repulsive forces become valid, the suspension is stabilized, and proper dispersion of a colloidal system is developed in this solvent. Experimentally, as milling time increased, the particle size in the suspension gradually changed from bimodal distribution to a nearly unimodal one.

4.3.9 CATALYSIS OF MECHANOCHEMICAL ORGANIC REACTIONS

Mechanical pretreatment of catalytic compositions is well documented (see, e.g., Kashkovsky 2003; Pauli and Poluboyarov 2003; Rac et al. 2005). This problem essentially falls into inorganic mechanochemistry and remains outside our consideration. Meanwhile, catalysis of organic mechanochemical reactions remains an unsolved problem. On the experimental level, it is possible to give some representative examples. These examples are representative in the sense that they demonstrate specificity of mechanically induced organic synthesis.

In the beginning, it is interesting to consider the typical catalytic reaction of carbon–carbon bond formation, namely, Suzuki coupling of boric acids with organic halides. The coupling proceeds in the presence of a palladium catalyst. In the mechanochemical variant of the reaction, it is most important, to obtain a good yield, to have good dispersion of the palladium complex on the potassium fluoride/alumina mixture. This dispersion is obtained by grinding the palladium catalyst with

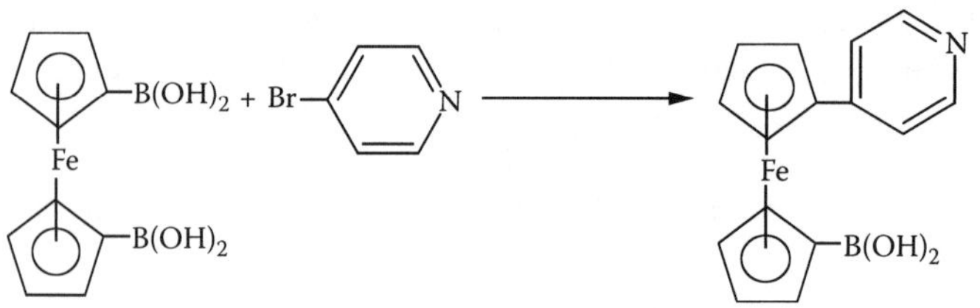

SCHEME 4.7

KF/Al_2O_3 before the reaction and, later, by adding a few drops of methanol to the mixture of Al_2O_3/reactants/catalyst. Methanol accelerates the solid-state process, transforming dry grinding into kneading. In particular, the reaction described was employed to prepare a pyridine derivative of ferrocenyl boric acid in air at ambient temperature. The product was obtained with a yield of about 60% according to Scheme 4.7 (Braga et al. 2004).

Dieckmann condensation provides another example of catalyzed mechanochemical reactions. If in a dibasic ester the hydrogen α to one ester group is δ or ε or $\sqrt{\varepsilon}$ to the other, intramolecular condensation may occur with the formation of a five- or six-member ring. This type of reaction, called *Dieckmann condensation*, is depicted in Scheme 4.8, in which diethyl adipate transforms into cyclopentanone-2-ethylcarboxylate under basic catalysis and in inert atmosphere. The reaction is usually carried out in solvent. High dilution is typical to avoid condensation between two molecules of diethyl adipate. Naturally, this brings about the ecological problem caused by large amounts of solvents.

In conditions of mechanochemical activation, the strictly intermolecular reaction proceeds in air and absolutely without solvents. The presence of potassium *tert*-butylate is essential for the reaction rate. Namely, after mixing of diethyl adipate and powdered potassium *tert*-butylate for 10 min at ambient temperature, the solidified reaction mixture was kept for 1 h to complete the interaction and to evaporate the alcohol formed. The dried reaction mixture was neutralized by addition of *p*-toluene sulfonic acid monohydrate and then distilled under reduced pressure to give cyclopentanone-2-ethylcarboxylate in a yield of more than 80% (Toda et al. 1998). This mechanochemical reaction represents a very clean, simple, environmentally friendly, and economical procedure.

The reaction between thiobarbituric acid and aromatic aldehydes can be performed on grinding, but it takes a long time (Li et al. 2002). The products are

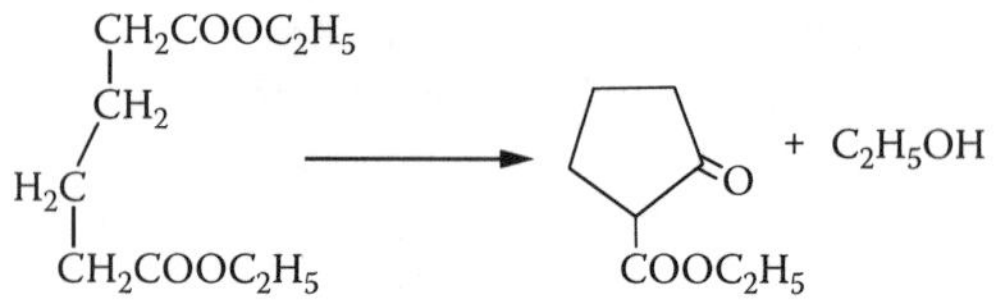

SCHEME 4.8

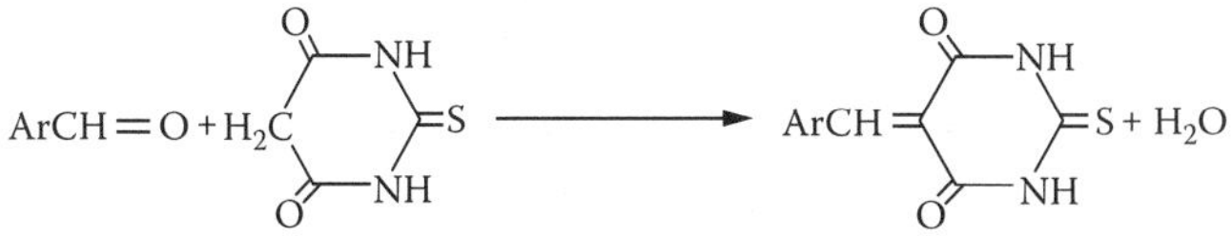

SCHEME 4.9

precursors for the synthesis of bioactive derivatives. In the presence of ammonium acetate, thiobarbituric acid and an aromatic aldehyde give rise to 5-arylmethyleneneth-iobarbituric acid in approximately 80% yield after 10 min of grinding in a mortar (Lu et al. 2004) (Scheme 4.9).

There are no explanations for the role of the additional reaction participants that accelerate the transformations of Scheme 4.9. In this sense, the work by Zhang et al. (2004) deserves special mention. The authors studied condensation of 5-(2-nitrophenyl)-2-furoyl chloride with ammonium thiocyanate and then with arylamine in the presence of poly(ethyleneglycol) (PEG-400). After about 5 min of mortar grinding, the final products were formed in almost quantitative yields (Scheme 4.10). The reaction proceeded through formation of the 5-(2-nitrophenyl)-furoyl isothiocyanate. PEG-400 acts as a catalyst that gives the complex (PEG-400/NH$_4$)$^+$SCN$^-$ with ammonium thiocyanate. Such complexation assists the total mechanochemical synthesis. Importantly, neither intermediate nor final compound forms in the absence of PEG-400.

Active centers in zeolite structures also exhibit their catalytic activity in dry synthesis. Such a feature found an application in the preparation of tris(isopropyl) borate. The ester is the starting material in manufacturing of boron single crystals for microelectronics. Boron anhydride and isopropanol are ground with zeolite spherical granules. In this case, the granules serve as milling solids and catalysts as well as water adsorbents. The method needs only 20 min to complete the esterification and is the method that consumer low energy (Molchanov, Buyanov, et al. 2002). The yield of the boron ester approaches 50%, whereas it barely reaches 25–30% with the conventional wet method.

Mechanical activation of crystalline anomers of D-glucose leaves them unaltered. However, mechanical activation in the presence of a solid acid (p-chlorobenzene

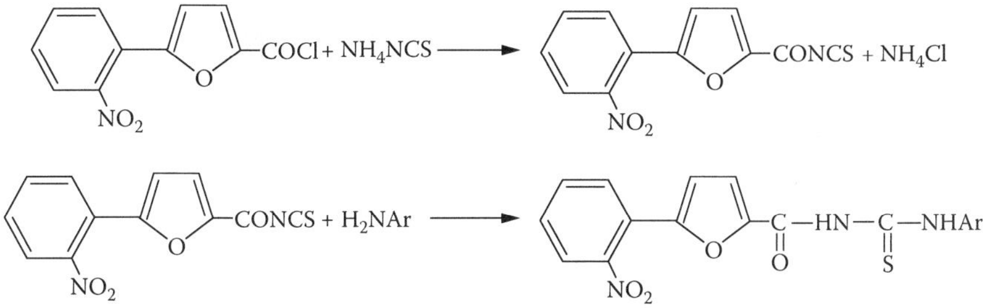

SCHEME 4.10

sulfonic acid) or a solid amphoteric electrolyte (sodium hydrogen carbonate) in amounts of only 2 wt% has brought about a transition of one anomer into the other. In contrast to sodium hydrogen carbonate, sodium carbonate (with no acidic hydrogen) was not active. Hence, acid catalysis of mutarotation takes place (Korolev et al. 2004). Mutarotation is a well-known process in carbohydrate chemistry. (Latin *mutare* means to change.) Intermittent addition of a proton to the pyranose oxygen leads to the open form of the glucose molecule. The open form then expels the added proton and undergoes cyclization into the other conformer. In the solid phase, a question arises on the way to accept and forward the proton in the course of mutarotation. It is obvious that the glucose molecules of the surface or at the surface are the proton acceptors. Moreover, the crystal structure of glucose has defects, as usual. The defects can also accept the catalyst (proton). It is the presence of the defect that makes it possible to forward the proton even deep into the depth of the crystal. As a result, the inversion proceeds in the sample mass.

4.4   MECHANOCHEMICAL APPROACHES TO FULLERENE REACTIVITY

Crossing of fullerene chemistry and friction physics is receiving attention in mechanical engineering. This crossing is expected to open a new technical field of molecular bearing that is promising for the realization of nano- and micromachines.

Generally, synthesis of fullerene derivatives is important to find new practical applications (see, e.g., Cao et al. 2002). However, poor solubility of fullerenes seriously restricts synthetic opportunities. Namely, the solubility of fullerenes in common organic solvents is so low that the use of a large amount of solvents becomes inevitable. Mechanochemical synthesis is mostly solvent free. Solvent-free reaction of fullerenes is an attractive and appealing method for synthesis of functionalized fullerenes. Mechanochemical solvent-free reactions of fullerenes have been developed (see references in Wang et al. 2003). The technique of high-speed vibration has been used to promote such reactions. In this technique, the mechanical energy caused by local high pressure, friction, shear strain, and the like can be transformed into driving force for the reaction. Such an advantage is illustrated next. However, one important note needs to be made before presenting these examples.

As seen in Chapter 3, fullerenes can be, principally, transformed into amorphous carbon on friction. In the sense of mechanosynthesis with fullerene participation, such a possibility presents some danger. Therefore, special experiments were undertaken to determine the survivability of crystalline C_{60} fullerene in the conditions of high-speed vibration milling. The most severe conditions were used for the test: vigorous motion along three axes with a frequency of 6000 r/min. At up to 30 min of such milling, no mechanical damage was observed in the C_{60} molecule. Even at milling for 5 h, this starting material remained intact for up to 65% (Braun et al. 2003). Usually, mechanochemical reactions are completed in 10–20 min and give rise to high enough yields of the desired products.

4.4.1 CYCLOADDITION

There is considerable interest in fullerene dimers. The unique physical properties of fullerene dimers offer potential access to novel molecular electronic devices. To date, however, accepted methods for synthesis of $(C_{60})_2$ and $(C_{70})_2$ dimers require high temperature or pressures as well as relatively complicated equipment. Mechanochemistry provides a chance to circumvent all of these obstacles. Moreover, the mechanochemical technique deals with solids. This is quite advantageous for the reaction of fullerenes, which are poorly soluble in common organic solvents.

Komatsu and coworkers (1998, 2000) reported preparation of $(C_{60})_2$ dimer from C_{60} via mechanochemical reaction using high-speed vibration milling in the presence of metals, potassium hydroxide, cyanate, thiocyanate, carbonate, acetate, or aminopyridines. The same dimer was also obtained by an even simpler manner, on grinding C_{60} fullerene in a mortar and pestle together with potassium carbonate (Forman et al. 2002). By the same method, $(C_{70})_2$ dimer was also prepared in this work from C_{70} monomer. In both reactions cited, [2 + 2]-type dimers formed. These dimers have two fullerene cages connected by a cyclobutane ring. Such dimers are the lowest energy and most plausible among possible isomers, including those with a peanut shell shape (see, i.e., calculations by Gal'pern et al. 1997 or Patchkovskii and Thiel 1998).

In vibration milling conditions, fullerenes react as electron acceptors when metals, salts, or amines step forward as electron donors (Komatsu et al. 2000). Particularly, the C_{60} fullerene molecule forms the radical anion $(C_{60})^-$ on action of potassium cyanide. Coupling with C_{60}, this radical anion gives $(C_{60})_2^-$. After that, a subsequent electron transfer to another C_{60} results in the formation of $(C_{60})_2$ and $(C_{60})^-$ as a new active species. Because of the lack of solvation on milling conditions, the radical anions formed are very active and react with the neutral surroundings. When the reaction between C_{60} fullerene and the same potassium cyanide is conducted in the dimethylformamide-dichlorobenzene mixture, the transformation takes a totally different course, and monomeric dicyanofullerene is formed (Keshavarz et al. 1995).

Because of the absence of any solvent molecule and the mechanical energy given to the reacting system, a mechanochemical reaction produces highly activated local sites in the reacting species. This reaction proceeds in heterogeneous solid-state conditions. Nevertheless, chemical equilibrium is established, starting from either monomer or dimer. Dissociation of the dimer is observed on mechanochemical activation (Komatsu et al. 1998, 2000).

Chemical equilibrium also characterizes [2 + 4] cycloaddition of anthracene to C_{60} fullerene under vibromilling (stainless steel balls in stainless steel capsule). In this example, the cycloadduct is formed at a 55% yield. Such yield is higher than that obtained in solution. This means that the equilibrium in the solid state lies more in favor of the [4 + 2] adduct than the reaction in solution (Murata et al. 1999; Wang et al. 2005). When the reaction of C_{60} fullerene with 9,10-dimethylanthracene was conducted in solution, it was impossible to isolate the corresponding cycloadduct because of the facile retro-cycloaddition. However, in the vibromilling experiments (30 min with immediate separation by flash chromatography on silica gel) the cycloadduct was isolated at a 62% yield. Although this adduct is stable in the solid state, it undergoes facile dissociation onto the initial compounds in solution at room temperature, with a half-life of about 2 h (Murata et al. 1999). This result clearly demonstrates the advantage of the solid-state reaction: It can lead to the formation of a thermodynamically unfavorable product.

4.4.2 FUNCTIONALIZATION

High-speed vibration milling has been successfully employed for functionalization of C_{60}. Introduction of naphthyl, phenyl, benzyl, adamantyl and fluorenyl substituents (Tanaka and Komatsu 1999) or an ester group (Wang et al. 1996) are representative examples. The solid-state reactions of substitution were tested using the corresponding organyl bromides in the presence of alkali metals or magnesium. Although yields of the substituted products are not high, two interesting features of the reaction deserve mention: (1) suppression of the fullerene dimerization or polymerization in the presence of the bromide, and (2) suppression of the Wurtz-type reaction in the presence of fullerene. (The Wurtz reaction leads to the formation of diorganyl products from organyl bromides in the presence of alkali metal.) Both peculiarities might be caused by the electron transfer mechanism of these solvent-free reactions. Sodium and lithium metals were as effective as potassium; magnesium was not so effective. In the absence of metal, no reaction was observed in the vibromilling treatment of C_{60} mixtures with organyl bromides. The problem of the solvent-free electron transfer (but in the immediate proximity of reacting particles) is still waiting detailed consideration.

Transformation of fullerenes into fullerols is another serious application of mechanosyntesis. Fullerols are one of the most interesting objects of fullerene chemistry. These water-soluble derivatives are attractive as a spherical molecular core in dendrimeric, star-shape polymers. Zhang et al. (2003) reported a solvent-free approach to C_{60} fullerols via simple solid-state reaction of C_{60} fullerene with potassium hydroxide under high-speed vibration milling. This approach needs no solvents; C_{60} fullerols are obtained in high yields, and the number of hydroxyl groups reaches 27. The reaction proceeds in air at room temperature. No unconsumed C_{60} fullerene or co-formed products are detected.

4.5 MECHANICALLY INDUCED REACTIONS OF PEPTIDES AND PROTEINS

The problem of mechanically induced reactions of peptides and proteins is relevant to phenomena occurring in living cells and organisms. Moreover, foodstuffs and feed products are prepared by mechanical processes that lead to destructive transformations of peptide moieties.

4.5.1 BOND RUPTURE

Mechanical treatment (grinding, stretching) gives rise to the formation of free radicals. In polypeptides, proteins (collagen, silk), globular proteins (trypsin, subtilisin, serum albumin), oligopeptides (gramicidin, bacitracin), the carbon-carbon bonds are cleaved predominantly:

$$—C(O)NHCH[R]C(O)NHCH[R]— \rightarrow —C(O)NHCH^{(\cdot)}[R] + {}^{(\cdot)}C(O)NHCH[R]—$$

This reaction leads to the initial terminal radicals (ITRs), which then react with polypeptide chains to give internal radicals (Dubinskaya 1999):

$$(ITR)^{(\cdot)} + —C(O)NHCH[R]C(O)NHCH[R]— \rightarrow$$

$$\rightarrow ITR–H + —C(O)NHC^{(\cdot)}[R]C(O)NHCH[R]—$$

With trypsin, albumin, and insulin, the mechanically induced cleavage of the polypeptide backbones is accompanied by rupture of the carbon–sulfur bonds (Dubinskaya 1999). The C—S bonds are weaker than the C—C bonds and rupture more rapidly. In the presence of atmospheric oxygen, carbon-centered radicals (both terminal and internal) are converted into peroxy radicals. Sulfur-centered radicals are stable and keep their integrity up to 340–360 K, at which temperature they become sufficiently mobile to recombine.

The rate of mechanical destruction and the rate of radical formation decrease with diminution in the molecular mass of the protein. Thus, the rate of radical formation from insulin (the species of the lowest molecular mass) is seven to nine times lower compared to the homolysis rates of other proteins (Dubinskaya 1999).

4.5.2 HYDROLYTIC DEPLETION

Mechanical dispersion of trypsin at 295 K led to the formation of glycine, tyrosine, serine, and glutamic and aspartic acids. When trypsin was dispersed at a low temperature (~80 K), amino acids were not formed (Yakusheva and Dubinskaya 1984).

4.5.3 BREAKAGE OF WEAK CONTACTS

Hydrogen bonds and hydrophobic and electrostatic interaction define the conformation and properties of polypeptides and proteins. In polymer chemistry, it is acknowledged that the mechanical response consists of breaking and reconnecting H-bonds under stress. However, the reconnection needs time (Shandryuk et al. 2003). Generally, the breaking of the weak contacts on mechanical stress causes conformational transition, disordering (loosening), and mechanical denaturation. For instance, collagen acquires the properties of flexible-chain gelatin (Dubinskaya et al. 1980). A water-soluble fraction has been found in the samples of globular proteins trypsin and subtilisin subjected to mechanical dispersion. The proportion of this fraction increases with grinding duration (Yakusheva and Dubinskaya 1984).

4.6 FORMATION OF MOLECULAR COMPLEXES

In some cases, mechanical treatment of binary mixtures leads to the formation of various molecular complexes. There are reports of mechanically induced formation of charge-transfer complexes and inclusion compounds as well as complexes formed because of acid–base interaction, hydrogen bonding, or simply van der Waals forces. Included in such complexes, medicinal drugs increase their therapeutic activity, which depends on bioavailability. Bioavailability of drugs poorly soluble in water is determined by the rate of dissolution. Milling of an active pharmaceutical ingredient is often employed to increase dissolution and promote homogeneity when the drug is mixed with fillers during preparation of tablets and the like. Of course, different drugs differ in their propensity for such fine mixing. For this reason, testing of different types of mills is required to establish eventually the right equipment for the drug form under preparation. Taylor and coauthors (2004) even proposed classifying drugs according to their brittleness indexes (and hence "millability"). The authors suggested that the index can be used to decide *a priori* which type of mill should be used.

Grinding of drugs with polymers of various structures leads to distribution (at the molecular level) of the drug in the polymer matrix to give molecular complexes. These systems are termed *solid dispersions*. As a rule, grinding of hydrophobic mixtures gives them solubility in water media at high enough dissolution rates. Meanwhile, by appropriate selection of the initial pair consisting of a polymeric carrier and a drug, the liberation of the drug from the molecular complex can also be retarded. This means that therapeutic action is prolonged. The polymeric matrices used for this purpose are usually natural polymers, namely, cellulose and derivatives, starch, cyclodextrins, proteins, chitin, chitosan, and pectin. Some synthetic polymers such as poly(ethylene oxide) or polyvinylpyrrolidone (PVP) are also used.

4.6.1 ACID–BASE COMPLEXATION

On grinding, a mixture of hexamethylenetetramine (commonly called urotropine) with resorcinol quantitatively transforms into the corresponding acid–base complex. Polymethacrylic acid and polymethylvinyltetrazol (a nitrogen base) also form a complex of such type under high pressure combined with shear stress (Dubinskaya 1999). The next example concerns co-grinding of silica with indomethacin [1-(4-chlorobenzoyl)-2-methyl-5-methoxyindole-3-acetic acid] (Watanabe et al. 2002, 2004a, 2004b). Such co-grinding is performed to prepare a solid dispersion with the enhanced dissolution rate. Mechanical stress exerted during co-grinding brings dissimilar particles closer to each other. Indomethacin drug acts as a Lewis base, and its carboxylic group is coordinated by the surface hydroxyl group of silica. Silica surface contains its own hydroxyls, each of them act as a Lewis acid. Such complexation holds the organic molecules at the silica surface, prevents the drug in the amorphous state from crystallization, and enhances the solubility and bioavailability of the drug.

4.6.2 CHARGE–TRANSFER COMPLEXATION

After grinding with additives (Ad), PVP gives the following types of charge-transfer complexes: PVP^+CA^- (CA = chloranil) or PVP^-PT^+ (PT = phenothiazine). Such complexation can sometimes develop further and lead to intermolecular electron transfer with the formation of radical ions (Dubinskaya 1999). Spectroscopy of electron paramagnetic resonance was used in studies of the samples obtained on triturating two substances in a mortar. Formation of stable radical ions was established (Tipikin et al. 1993). The authors checked phenol, catechols, and porphyrins as electron donors and quinones as electron acceptors. According to the electron spectra, the primary reaction consisted of the formation of triplet complexes or ion–radical pairs. The next stage was the formation of free-radical ions. The triplet molecular complexes or ion–radical pairs generated by mechanical treatment are much more stable than those prepared by photolysis of the same donor–acceptor solid mixture. Apparently, once formed, paramagnetic species are quickly incorporated into the new crystal lattice arising on mechanical treatment.

A detailed study of mechanochemical stabilization of the paramagnetic species was undertaken (Tipikin 2002). An electron paramagnetic resonance spectrum was recorded during co-joint grinding of oxalic acid and urea. The spectrum was kept during several days for the ground mixture in the solid state. Additional grinding regenerated the spectrum when it disappeared. Dissolution of the sample led to annihilation of the paramagnetic particles. Similarly, grinding of azo-bis(isobutyronitrile) in the presence of ballast compounds — oxalic acid and m-nitroaniline — gave rise to stabilized solid-state butyric radicals. The radicals were not observed without the ballast.

4.6.3 HOST–GUEST COMPLEXATION

Formations of host–guest complexes were reported in cases of mechanical grinding of cyclodextrins with benzoic acid derivatives (Nakai et al. 1984). Cyclodextrins are cyclic oligosaccharides that possess a cavity capable of forming host–guest, or inclusion, complexes with a variety of organic molecules. The diameter of the cyclodextrin cavity is a function of the number of glucose residues that form the cavity inner wall.

Mechanochemically induced inclusion reactions lead to the entropy-frozen systems. This improves aqueous solubility and oxidation stability of many practically important compounds. Thus, mechanical activation of the mixture containing bioactive compounds and filling materials finds wide application in pharmacy. For instance, carotene, coumarin, riboflavin derivatives, and vitamins A, E, and K are soluble only in oils. Their inclusion reactions with dextran, dextrin, or PVP lead to water-soluble complexes. The constituents of these complexes are not cleaved and do not lose their biological properties under mechanical activation (Chuev et al. 1991).

Italian chemists at Carlo Erba Pharmaceutical Company compared physical and pharmaceutical properties of one drug (methyl hydroxyprogesterone acetate) obtained by co-precipitation or co-grinding with β-cyclodextrin. The co-precipitated form presented a coarser particle size distribution; the co-ground form had smaller dimensions. The pure drug has very slight aqueous solubility. The drug inclusion into dextrin enhanced the solubility and, in the case of co-grinding, at a much higher degree than for co-precipitation. In dogs, the mechanically activated system gave the highest progesterone blood level, 6-fold greater than tablets of the mixture and 2.5-fold greater than for the co-precipitated system (Carli et al. 1987).

Although fullerenes are nonpolar molecules, there have been various attempts to make them water soluble in view of their applications in the biomedical field. In particular, fullerene inclusion in cyclodextrins can lead to water-soluble complexes. According to Andersson et al. (1994), C_{60} fullerene does not react with β-cyclodextrin. However, Murthy and Geckler (2001, 2002) later obtained such inclusion complexes.

The data on the dimensions of the α-, β-, and γ-cyclodextrins were considered in detail (see Murthy and Geckler 2002 and references therein). The dimensions of the cyclodextrins do not rule out the formation of their complexes with C_{60} fullerene. In these complexes, the fullerene molecule does not intrude deeply into the dextrin cavity. For β-cyclodextrin, the cavity diameter is 780 pm, and the outer rim diameter is 1530 pm (on the polar side of the molecule) compared to 950 pm and 1690 pm, respectively, for γ-cyclodextrin. Therefore, an inclusion compound is formed in which one C_{60} fullerene is "encompassed" by two cyclodextrins.

The C_{70} fullerene size is greater than that of C_{60} fullerene. Nevertheless, both C_{60} and C_{70} pieces were included in the γ-cyclodextrin cavities on mechanochemical initiation (Braun et al. 1994; Komatsu et al. 1999). Unfortunately, the authors did not mention whether 1:1 or 1:2 complexes were formed. According to Braun et al. (1995), the reaction presumably proceeded via the mechanochemical amorphization of cyclodextrin. Fullerene is also rendered amorphous, although to a lesser degree. After that, dissolution of a fullerene in the amorphous phase of γ-cyclodextrin takes place under vigorous mechanochemical treatment. The phase transition is stopped when milling is discontinued.

If the guest molecule is bulky enough, cyclodextrin can include only a part of it. Thus, when the free-radical probe α-phenyl-α-(2,4,6-triomethoxybenzyl) *tert*-butyl nitroxide $[(CH_3)_3C_6H_2–CH(C_6H_5)N(O^.)\text{-}tert\text{-}C_4H_9]$ was mixed with γ-cyclodextrin in a water solution at ambient pressure, two different, phenyl-in and *tert*-butyl-in, complexes were identified by ESR. With increasing external pressure, the equilibrium between the *tert*-butyl-in and phenyl-in complexes shifted to the phenyl-in complex side. In contrast, when β-cyclodextrin was used, the equilibrium shifted to the *tert*-butyl-in complex side. There is clear correlation between sizes of the host cavity and the fragmental volume of the group inserting. Pressure forces the bulkier group to drive in the cavity and shifts the equilibrium to the thermodynamically unfavorable side (Sueishi et al. 2004).

Host–guest complexation is sensitive to temperature. For instance, the inclusion reaction between ursodeoxycholic acid (one of the bile acids) and phenanthrene is completed by conventional grinding at ambient temperature. Grinding at lower temperatures (external cooling with ice, dry ice, or cold nitrogen gas) provides a mixture of the amorphous acid and finely crystalline phenanthrene (Oguchi et al. 2003). No complex is formed. Obviously, disintegration and heating function cooperatively for the formation of the host–guest complex under consideration. The temperature helps the physical mixture to reach the lower energy level needed for the inclusion reaction.

4.6.4 FORMATION OF HYDROGEN-BONDED AND VAN DER WAALS COMPLEXES

In 1983, Ioffe and Ginzburg showed that cholesterol forms van der Waals complexes with carboxylic acids or their amides. These complexes are weak (their formation enthalpies do not exceed 30 kJ/mol). Nevertheless, van der Waals forces assist extraction of plant bioactive substances during grinding of raw materials in the presence of a solid but water-soluble collector (e.g., saccharose). Intensive mechanical treatment disturbs cell shells. Their content comes in close contact with the solid collector. Physical factors of the process are also important for the extraction considered.

At intense mechanical activation, local zones appear where temperature and pressure are critical. For moist-mixture processing, porous water transfers into a supercritical state. This strongly enforces the dissolving power of the water, which also helps transport solutes. Formation of a steady solution of molecular complexes results from water treatment of the mixture obtained mechanochemically. Thus, from the raw herb material *Serratula coronata,* phytoecdysteroids were efficiently isolated as a dry powder (Lomovskii et al. 2003). The dry powder was used as a fodder component to test its estrogenic activity. Within the first three months after accouchement the tested female animals restored their normal ovulation. After fertilization, breeding productivity, was 30% more than the control group. In other tests, the phytosteroid additive boosts protein biosynthesis in animal liver, kidneys, and muscular tissues. This property is widely used to enhance the physical capabilities of professional sporting persons. The Lomovskii et al. stress that usage of the phytoecdysteriods is not accompanied with dangerous consequences for life as distinct from synthetic steriods.

Mechanical treatment favors hydrogen bonding between participants of solid-state reactions. Such a method is efficient for preparing a wide variety of hydrogen-bonded organic cocrystals, particularly when one component is a good proton donor and the other is a good proton acceptor.

For carboxylic acid pairs, solid-state cocrystal formation involving heterodimeric association is favored when the two acids have different acidities. Thus, the hydrogen-bonded complex is quantitatively formed by grinding of 4-choro-3,5-dinitrobenzoic acid and 4-aminobenzoic acid in a mill for about 20 min at room temperature. Heating of this complex leads to the formation of the nucleophilic aromatic substitution product according to Scheme 4.11 (Etter et al. 1989).

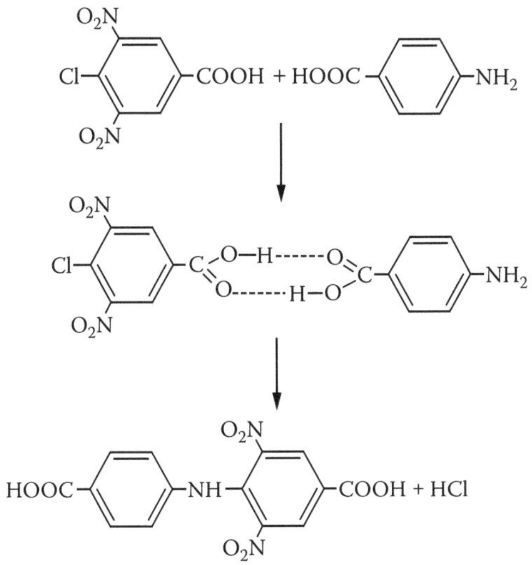

SCHEME 4.11

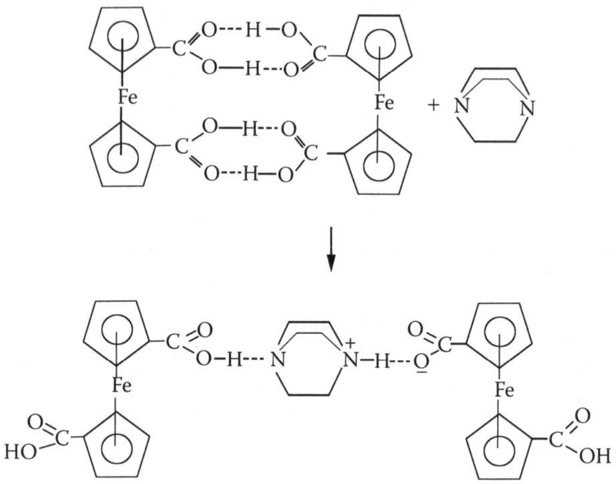

SCHEME 4.12

Manual (pestle-and-mortar) grinding of solid ferrocenyl dicarboxylic acid and 1,4-diazabicyclo[2.2.2]octane gives rise to quantitative formation of a salt with counterions also connected by hydrogen bonds (Braga et al. 2002; Braga, Maini, de Sanctis, et al. 2003; Braga, Maini, Polito, et al. 2003). The reagents form a hybrid organic-organometallic species. The following three events emerge co-jointly: (1) acid–base interaction (proton transfer) from the acid to amine; (2) *cis-trans* transition of the 1′,1″-disubstituted ferrocene; and (3) formation of the ternary hydrogen-bonded complex. Scheme 4.12 illustrates all of the events.

Scheme 4.12 shows that the diazacyclooctane cation plays the role of a bridge connecting the two sandwich molecules in the *trans* form. The starting crystalline ferrocenyl dicarboxylic acid exists as the hydrogen-bonded dimer in which the carboxylic groups are, by necessity, located in the *cis* position in relation to each other (Palenic 1969).

In the case of malonic acid (in which the two carboxylic groups are not so distant), both hydrogen bonds — intramolecular and intermolecular — are established as a result of manual grinding (Scheme 4.13) (Braga, Maini, de Sanctis, et al. 2003).

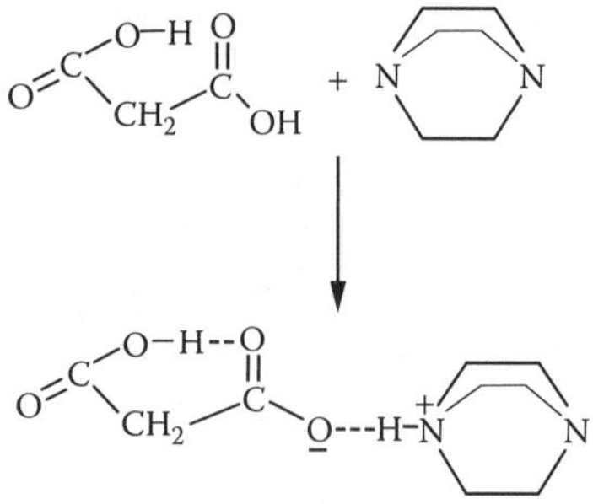

SCHEME 4.13

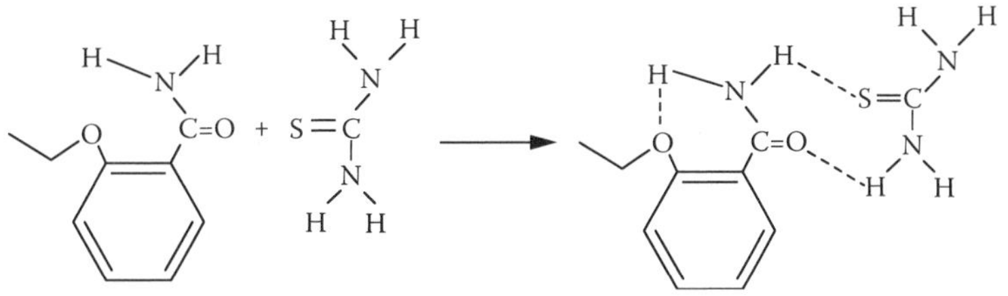

SCHEME 4.14

Grinding of thiourea and ortho-ethoxybenzamide (ethenzamide drug) leads to a hydrogen bond complex with a structure established by powder x-ray diffraction and depicted in Scheme 4.14 (Moribe et al. 2004). Formed during mechanical treatment, this complex contains a C=O···H—N intramolecular bond and N—H···S=C and C=O···H-N intermolecular bonds. It is worth noting that solid thiourea (introduced in the reaction in Scheme 4.14) contained N—H···S=C intramolecular bonds. Consequently, formation of the resulting complex is accompanied by the replacement of one intermolecular hydrogen bond with another.

The majority of drugs are low molecular mass organic compounds containing functional groups capable of forming intermolecular hydrogen bonds. Carbonyl, hydroxy, amino/imino, and other similar groups are typical moieties of drugs. During mechanical grinding of mixtures of drugs with polymers, the intermolecular hydrogen bonds are destroyed, and new hydrogen bonds with macromolecules are formed. Dubinskaya (1999) reviewed evidence of such weak complexation for several examples. Benzoic, salicylic, acetylsalicylic, and other acids form hydrogen bonds of the C=O···HOR type with cellulose and oligosaccharide. Derivatives of barbituric acid form NH···O(H)R bonds with hydroxy groups of polymers. In the inclusion compounds obtained by grinding of cyclodextrins with acetylsalicylic acid (aspirin) and benzoic and p-hydroxybenzoic acids, hydrogen bonds link the OH groups of dextrins to the C=O groups of acids. On grinding of ibuprofen (isobutyl phenyl propionic acid) with poly(ethyleneglycol), the hydrogen bond between the carboxylic group of the acid and the hydroxy group of the polymer were detected, as were van der Waals interactions between the polymer molecules and the aromatic ring of ibuprofen.

As stated, enhancement of the dissolution rates of poorly water-soluble compounds can expedite the process of formulation design. Amorphization of drugs increases their dissolution, which in turn increases their bioavailability. Conversion to the amorphous state of a drug is, of course, desirable. Often, however, reversion from the amorphous to the lower energy crystalline state is observed. Reversion has been a major limitation in the successful commercialization of solid dispersions, an approach to enhance dissolution of poorly water-soluble drugs. There is therefore a need to stabilize the resulting amorphous state.

The pharmaceutical adsorbent Neusilin was used to stabilize the amorphous states of such drugs as ketoprofen, indomethacin, naproxen, and progesterone (Gupta et al. 2003). Neusilin consists of amorphous microporous granules of magnesium alumosilicate ($MgO \cdot Al_2O_3 \cdot SiO_2$) with a high specific surface area (~ 300 m^2 g^{-1}).

Neusilin has silanol groups on its surface along with metal oxides. In the work cited, each of the four enumerated drugs was milled with Neusilin to effect conversion from crystalline to amorphous states, and the physical stability of the resultant drugs was studied. Ball milling the drugs alone for 48 h did not result in amorphization. In the presence of Neusilin, ball milling did lead to amorphization. Whereas the carboxylic acid-containing drugs (ketopofen, naproxen, and indomethacin) interact with Neusilin via acid–base interaction, hydrogen bonding is responsible for the interaction of progesterone with Neusilin. Progesterone bears no carboxylic group but has a carbonyl group. It is this group that acts as a proton acceptor and forms a hydrogen bond with a surface hydroxyl group of Neusilin.

It needs to be emphasized that the interaction considered above between the drugs bearing a carboxylic group and Neusilin also begins from hydrogen bonding. After that, electrostatic forces are established between COO^- and counterions such as Mg^{2+} and Al^{3+}. These electrostatic forces and hydrogen-bonding interactions drive the irreversible amorphization of the drugs. The amorphous Neusilin-bound states of all four drugs are stable during storage.

4.7 MECHANICAL INITIATION OF INTERMOLECULAR ELECTRON TRANSFER AND INTRAMOLECULAR ELECTRON REDISTRIBUTION

To illustrate mechanical initiation of electron transfer, the reaction of alkyl halides with metallic aluminum should be cited (Mori et al. 1982). No reaction of unmilled aluminum powder with alkyl halides was observed during 10 h of contact. When aluminum was milled with stainless steel balls in a stainless steel pot under helium at room temperature in the presence of butyl iodide for 8 min, an exothermic reaction was initiated, and no additional activation was required to move the reaction forward. If additional butyl iodide was injected into the mixture, the reaction continued without milling until aluminum was exhausted. Little gaseous product was evolved. The distillate of the liquid product was colorless.

Nuclear magnetic resonance measurements confirmed that the carbon–aluminum bond did exist in the distillate. When the distillate was hydrolyzed with water, butane evolved. The amount of butane was nearly equivalent to that of the reacted butyl iodide. The equivalent amount of iodide ion was detected in aqueous solution. From the results, the liquid product was identified as butyl aluminum iodide:

$$3C_4H_9I + 2Al \rightarrow (C_4H_9)_3Al_2I_3$$

As mentioned, once the reaction was initiated by mechanical activation, it was continued autocatalytically. Regarding the fate of butyl bromide, 1.1 mmol of a gaseous product was evolved for the reaction of the starting C_4H_9I (4.6 mmol) after 12 min milling. The main component of that gas was butane. Butene was a minor product. The residue was a viscous, dark brown material. Bromide ion (4.6 equivalents) was detected in the residue. The residue was a mixture of aluminum bromide and polymerized matter. This vibromilling reaction may be described as follows:

$$3C_4H_9Br + Al \rightarrow AlBr_3 + 3(C_4H_9^{\cdot})$$

$$2(C_4H_9^{\cdot}) \rightarrow C_4H_{10} + C_4H_8$$

$$C_4H_8 \rightarrow Polymer$$

The mechanochemical reaction of aluminum with butyl bromide was investigated under two reaction conditions: during and after milling. The active source may be different in the two reactions. However, high temperature, high pressure, and nascent surface did not appear to be active factors in this case because preactivated aluminum was observed to react with butyl bromide even after the termination of milling.

Obviously, the lattice disorder and the Kramer effect remains to be analyzed. An x-ray study showed that the lattice disorder in aluminum increased slightly when milled and did not change with time. Consequently, the lattice disorder is not the main cause of the mechanochemical activity. In the meantime, the reactivity of milled aluminum correlated well with the intensity of the exoelectron emission. Such an emission gradually decayed after termination of milling, along with the suppression of the chemical reaction. The aluminum, which had entirely lost electron emission activity, did not react with butyl bromide.

The electron emission intensity of the free (unused) electrons under butyl bromide "atmosphere" was less than 20% of that under benzene atmosphere. In other words, exoelectrons are better captured with butyl bromide than with benzene. Butyl bromide has much more electron affinity than benzene. In the process considered, butyl bromide captured free electrons. It is clear that the exoelectrons resulting from aluminum vibromilling initiate the reaction between aluminum and organic acceptors.

Complexation of 1,10-phenanthroline (phen) to iron pentacarbonyl [Fe(CO)$_5$] represents a special case of the mechanical activation. After water treatment of [phen + Fe(CO)$_5$] ground mixture, the complex [Fe(phen)$_3$]$^{2+}$(phen)$^{-\cdot}$(HO)$^{-}$·3H$_2$O was isolated (Drozdova et al. 2003). The reaction was performed in a sealed vibration ball mill, all elements of which were stainless steel. According to the present consideration, we have to pay attention to (phen)$^{-\cdot}$ formation. This is a result of mechanically induced electron transfer. It is only unclear, however, why this radical anion remains stable in air and resistant to protonation from the neighboring water as a proton donor.

An intriguing redox reaction has been found between metal oxides and quinones under mechanical pulse action (Aleksandrov et al. 1999). The following metal oxides were successfully tested: CuO, ZnO, CdO, PbO, Al$_2$O$_3$, Ga$_2$O$_3$, Sb$_2$O$_3$, Bi$_2$O$_3$, Cr$_2$O$_3$, TiO$_2$, GeO$_2$, ZrO$_2$, or SnO$_2$. The metal-containing radical anions of quinones were formed both on the surface of metal oxides and as individual solid phases. In the solid-state mixtures, these radical anions are stable for months. During the mechanochemical action, metal oxide surface is approximately 70% transformed into the active state. This surface is in the zone of action of elastic distortions about dislocations that reach the surface. Therefore, coordinately unsaturated metal ions appear when dislocations reach the oxide surface. The metal ions on the surface are

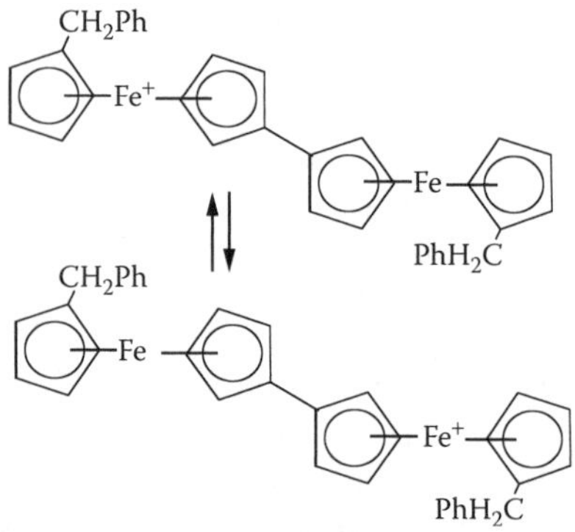

SCHEME 4.15

transformed to the zero-valence state through reduction reactions at the broken metal–oxygen bonds. Such breakage occurs when dislocations reach the surface. The zero-valence metals are obviously responsible for the transformation of quinone molecules into semiquinones (quinone radical anions). The latter are bound in complexes with metal cations. To sum, metal oxides step forward in an unusual role of "electron donors."

Scheme 4.15 describes the effect of grinding on the rate of intramolecular electron transfer in organic crystals. Namely, grinding of $1',1''$-dibenzylbiferrocenium iodide needle crystallite results in deceleration of the reaction in Scheme 4.15 by three orders of magnitude (Webb et al. 1991).

Mechanical grinding is known to increase the number of defects in a crystalline lattice. Because it is proximal to a defect, $1',1''$-dibenzylbiferrocenium is trapped, and its Moessbauer spectrum is drastically changed. Physically, trapping shifts the zero-point energy level of the ferrocenium derivative so that intramolecular electron transfer is slowed as a result of grinding. (Nevertheless, there is no marked effect on the x-ray diffraction patterns of the samples before and after grinding.) Slowing of intramolecular electron transfer was also observed for $1',2',1'',2''$ (or $1',3',1'',3''$)-tetra(naphthylmethyl) biferrocenium iodide (Dong et al. 2003). So, mechanical action affects intramolecular electron redistribution. Practical applications must be developed for this subtle phenomenon.

A special aspect of intramolecular electron transfer is the metal–insulator transition induced by application of pressure. At 10-MPa pressure, copper complex bearing two cation-radical-salt ligands undergoes so-called Mott transition coupled with the Peierls transition. The ligand was a dimethyl derivative of N,N'-dicyanoquinonediimine cation–radical salt with iodide anion. The applied pressure affects the coordination geometry around copper and regulates the degree of electron transfer from the metal to the ligand (Kato 2004).

Mechanical action of a shock wave can initiate electron redistribution. The following example is demonstrative: Hexogen {cyclo-$CH_2 N(NO_2)]_3$}, which also is called Cyclonite, RDX, or 1,3,5-trinitro-2,2,4,4,6,6-hexahydro-1,3,5-triazine, is an important secondary (brisant) explosive with wide application in warfare, in rock ammonite, and for shooting oil wells through. Its detonation is induced by shock waves. The initial step of molecular decomposition was considered by many authors; for a list of references see the work of Chakraborty et al. (2000). As suggested, the decomposition can principally proceed through the formation of radicals $CH_2N_2O_2$, NO, HONO, or NO_2. The problem was carefully analyzed by Luty and colleagues (2002) by density functional theory.

The shock wave provides rapid input of mechanical energy. A compound suffering such an input undergoes excitation, and a sudden change in the electronic distribution takes place. Namely, an electron jumps from the highest occupied molecular orbital (HOMO) onto the lowest unoccupied molecular orbital (LUMO). It means narrowing of the gap between HOMO and LUMO. The smaller the gap is, the more readily a unimolecular reaction can occur. When the molecule is driven to critical electronic excitation, the new state is formed. This state has equally occupied HOMO and LUMO. It is equal to closure of the gap. If the gap closes completely, the molecule will decompose instantaneously, and Hexogen detonation occurs.

In nitro compounds, the LUMO is preferentially localized in the framework of the nitro group. Consequently, the most probable, mechanically (shock-wave) induced, unimolecular reaction in Hexogen is elimination of the nitro group with breaking of an $N-NO_2$ bond. And, only NO_2 is eliminated although the gas-phase energy barrier for HOMO elimination is essentially the same (Strachan et al. 2003).

In definite circumstances, mechanical grinding may cause one-electron oxidation that enhances activity of solid dugs. For instance, chemotherapy's use of officinal adriamycin (doxorubicin hydrochloride) had no effect in adriamycin-resistant Guerin's carcinoma. However, administration of the mechanically modified drug resulted in $52 \pm 4\%$ decrease of such tumor volumes in animals compared to the control group and the group treated with officinal adriamycin. According to proton magnetic resonance studies, mechanically activated adriamycin contains an increased concentration of monovalent and divalent positively charged ions (Todor et al. 2002). It is likely that mechanical treatment results in generation of a positively charged oxygen-containing group (say CH_3O^+-) that complements the ammonium moiety (Scheme 4.16).

The course of events (including generation of superoxide ion and hydroxyl radical as a one-electron oxidant) is described in Chapter 3 (Section 3.2). In Chapter 3, transition of alkenes into alkane carboxylic acids is also described for friction in the presence of oxygen and water. The same transition has been found in mechanically induced oxidation of alkenes by potassium permanganate. In this reaction also, the presence of water enhances product yield (Nuchter et al. 2000).

4.8 MECHANICALLY INDUCED CONFORMATIONAL TRANSITION OF ORGANIC COMPOUNDS

Conformational composition is very important in pharmaceutical manufacturing. In this field, mechanical treatment of drugs and their ready-made forms are increasingly used. Understandably, the available data should be considered in this chapter. Many reports revealed that ignoring this issue can lead to a useless drug product or even toxicity. Because many drugs are administered as racemates, racemic modification in the solid crystalline state is a significant problem.

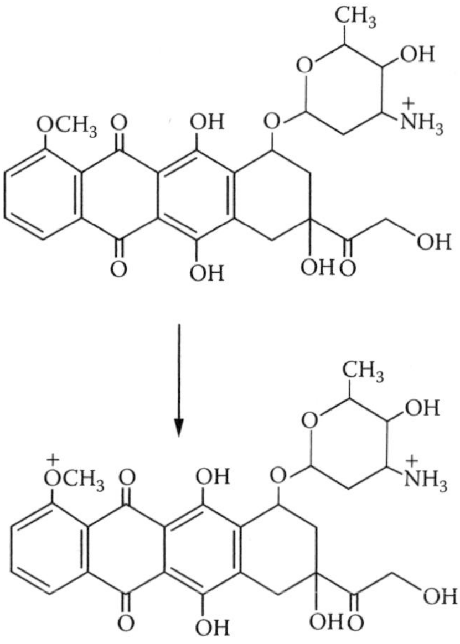

SCHEME 4.16

A *racemate* is defined as an equimolar mixture of two enantiomers that can belong to one of three different classes depending on its crystalline arrangement. The first is a *racemic compound,* in which the two enantiomers are present in equal quantity in a well-defined arrangement within the same unit cell. The second class of a crystalline racemate is a racemic mixture or *conglomerate,* that is, a separable mechanical mixture of the two pure enantiomers. The third class, called the *pseudoracemate* or the *racemic solid solution,* is a solid formed by the two enantiomers coexisting in an unordered manner in the crystal.

It is interesting to track the effects of various grinding regimes on racemization of amino acids. For leucine, norleucine, valine, or serine, physical mixing of equimolar quantities of D and L crystals using mortar and pestle results in formation of conglomerate. Grinding the physical mixture of each amino acid enantiomer with a vibration mill led to formation of a racemic compound. The conglomerates and racemic compounds are clearly distinguished by powder x-ray diffraction, differential scanning calorimetry, and infrared spectroscopy (Piyarom et al. 1997). These authors did not report the fate of D or L amino acid on grinding each entaniomer separately.

Ikekawa and coauthors (Ikekawa and Hayakawa 1991a, 1991b; Ikekawa et al. 1990) did fulfill such work as they studied racemization of L-phenylalanine in grinding conditions. The amino acid was ball milled with various inorganic powders (kaolin, talc, silica, alumina, or magnesia). The surface area of inorganic powders ball milled together with L-phenylalanine was much greater than the area of the inorganic powders ball milled separately without phenylalanine. As suggested, the inorganic surface was coated with thin films of phenylalanine in a ball-milled mixture. This supposedly prevents fine inorganic particles from aggregation or agglomeration. Having been ground together with an inorganic component, L-phenylalanine undergoes partial racemization that changes the infrared spectral picture of the amino acid.

The following mechanism of racemization obviously occurs: An inorganic fine powder acquires large surface energy and tightly absorbs L-phenylalanine. As a result, the latter is distorted. The distorted and highly activated state of phenylalanine is stabilized through conformational transition. The excess energy is sufficient for the conformational change of part of the molecules. That is why the L-enantiomer racemization is observed, leading to the mixture of L- and D-phenylalanines.

Interestingly, when the D,L-valine crystals were ground at low temperature, the one-dimensional expansions of the crystal lattice along the b-axis occurred concurrently with the formation of an amorphous component. In the $(CH_3)_2CHCH$ $(NH_3^+)COO^-$ ensemble, intermolecular hydrogen bonds between NH_3^+ and COO^- groups are changed in the mode. The infrared spectra and x-ray diffraction pattern testify that grinding causes defects in valine crystals, particularly in the hydrogen bond layers. This results in an increase in crystal volume (Moribe et al. 2003).

Sometimes, grinding leads to conformational changes in only part of a molecular framework. This is detectable according to changes in physicochemical properties of the crystalline material. Thus, by grinding crystalline ferrous tris(1,10-phenanthroline) bis(hexafluorophosphate) $\{[Fe^{II}(phen)_3](PF_6)_2\}$, the effective magnetic moment μ_{eff} increases, and simultaneous amorphization takes place. Subsequent annealing further increases μ_{eff} despite restoration of the crystallinity. The authors (Ohshita et al. 2004) explained the phenomenon as the recovery of the large counterion PF_6^- from its strained state in the intact crystal to a less-strained state toward higher spherical symmetry. Restoration of the spherical symmetry of PF_6^- induced an increase in the free volume and delocalization of electron density around a phen ring. This reduces the ligand field strength and hence increases the effective magnetic moment.

4.9 CONCLUSION

The examples considered deal with utilization of mechanical energy as a driving force for chemical reactions. Here, the generation of local high-pressure spots is presumed to activate local reaction sites. Furthermore, the absence of any solvent molecule brings the reacting species into the closest contact without any solvation. Such a reaction system causes a novel chemical reaction to occur. Of course, the main targets of our attention were the reactions leading to new desired products.

Chapter 4 describes examples indicating that solvent-free mechanochemistry has the potential to become an alternative to conventional organic synthesis. Continuing research to establish the suitability of mechanochemical synthesis to different reaction types may lead to the development of novel green chemistry. Of course, solvent-free mechanochemistry is an energy-intensive technique compared with conventional solvent-based chemical synthesis. However, the energy needed to produce, deliver, collect, and dispose of the solvents and restore the environment is considerably higher. Therefore, the advantages of the mechanochemical approach are noteworthy.

REFERENCES

Acheson, R.M., Matsumoto, E.K. (1991) *Organic Synthesis at High Pressure* (Wiley, New York, 1991).

Akhmetkhanov, R.M., Kadyrov, R.G., Kolesov, S.V. (2004) *Khim. Promyst. Segodnya,* **9**, 41.

Aleksandrov, A.I., Prokof'ev, A.I., Bubnov, N.N., Rakhimov, R.R., Aleksandrov, I.A., Dubinskii, A.A., Lebedev, Ya.S. (1999) *Izv. Akad. Nauk, Ser. Khim.,* 324.

Aleksandrov, A.I., Zelenetskii, A.N., Krasovskii, V.G., Aleksandrov, I.A., Makarov, K.N., Prokof'ev, A.I., Bubnov, N.N., Dobryakov, S.N., Rakhimov, R.R. (2003) *Dokl. Akad. Nauk* **389**, 640.

Andersson, T., Westman, G., Wennerstrom, O., Sundahl, M. (1994) *J. Chem. Soc., Perkin Trans.* **2**, 1097.

Balema, V.P., Wiench, J.W., Pruski, M., Pecharsky, V.K. (2002a) *Chem. Commun. (Cambridge, UK)*, 724.

Balema, V.P., Wiench, J.W., Pruski, M., Pecharsky, V.K. (2002b) *Chem. Commun. (Cambridge, UK)*, 1606.

Balema, V.P., Wiench, J.W., Pruski, M., Pecharsky, V.K. (2002c) *J. Am. Chem. Soc.* **124**, 6244.

Boehme, K., Weber, M., Schmidt-Naake, G. (2003) *Chem. Ing. Tech.* **75**, 249.

Boldyrev, V.V. In: *Reactivity of Solids: Past, Present and Future*, Edited by Boldyrev, V.V. (Blackwell Science, Oxford, UK, 1996, p. 267).

Borisov, A.P., Bravaya, N.M., Makhaev, V.D. (2004) *Russ. Pat.* 0223067.

Borowitz, J., Rusek, P.E., Virkhaus, R. (1969) *J. Org. Chem.* **34**, 1995.

Braga, D., D'Addario, D., Polito, M., Grepioni, F. (2004) *Organometallics* **23**, 2810.

Braga, D., Maini, L., Giaffreda, S.L. Grepioni, F., Chierotti, M.R., Gobetto, R. (2004) *Chem. Eur. J.* **10**, 3261.

Braga, D., Maini, L., Polito, M., Mirolo, L., Grepioni, F. (2002) *Chem. Commun. (Cambridge, UK)*, 2960.

Braga, D., Maini, L., de Sanctis, G., Rubini, K., Grepioni, F., Chierotti, M.R., Gobetto, R. (2003) *Chem. Eur. J.* **9**, 5538.

Braga, D., Maini, L., Polito, M., Mirolo, L., Grepioni, F. (2003) *Chem. Eur. J.* **9**, 4362.

Braun, T., Buvari-Barcza, A., Barcza, L., Konkoly-Thege, I., Fodor, M., Migali, B. (1994) *Solid State Ionics* **74**, 47.

Braun, T., Buvari-Barcza, A., Barcza, L., Konkoly-Thege, I., Fodor, M., Migali, B. (1995) *Magy. Kem. Foly* **101**, 76.

Braun, T., Rausch, H., Biro, L.P., Zsoldos, E., Ohmacht, R., Mark, L. (2003) *Chem. Phys. Lett.* **375**, 522.

Briand, G.G., Burford, N. (1999) *Chem. Rev.* **99**, 2601.

Cao, T., Wei, F., Yang, Y., Huang, L., Zhao, X., Cao, W. (2002) *Langmuir* **18**, 5186.

Carli, F., Colombo, I., Torricelli, C. (1987) *Chim. Oggi*, **5**, 61.

Chakraborty, D., Muller, R.P., Dasgupta, S., Goddard, W.A. (2000) *J. Phys. Chem. A* **104**, 2261.

Chetverikov, K.G., Arkhireev, V.P., Kochnev, A.M., Mutriskov, A.P. (2002) *Izv. Vysshikh Uchebnykh Zavedenii, Khim. Khim. Tekhnol.* **45**, 121.

Christe, K.O., Dixon, D.A., Mc Hemore, D., Wilson, W.W., Sheehy, J.A, Boatz, J.A. (2000) *J. Fluorine Chem.* **101**, 151.

Chuev, V.P., Kameneva, O.D., Chikalo, T.M., Nikitchenko, V.M. (1991) *Sib. Khim. Zh.* 156.

Chuev, V.P., Lyagina, L.A., Ivanov, E.Y., Boldyrev, V.V. (1989) *Dokl. Akad. Nauk SSSR* **307**, 1429.

Dong, T.Y., Lin, M.-Ch., Lee, L., Cheng, Ch.-H., Peng, Sh.-M., Lee, G.-H. (2003) *J. Organometal. Chem.* **679**, 181.

Drozdova, M.K., Myakishev, K.G., Ikorskii, V.N., Aksenov, V.K., Volkov, V.V. (2003) *Koordinatsionnaya Khim.* **29**, 250.

Du, B., Du, Ya., Wang, L., Wang, M., Gao, H., Dong, L. (2002) *Shipin Gongye Keji* **23**, 38.

Dubinskaya, A.M. (1999) *Uspekhi Khim.* **68**, 708.

Dubinskaya, A.M., Segalova, N.E., Belavtseva, E.M., Kabanova, T.A., Istranov, L.P. (1980) *Biofizika* **25**, 610.

Dushkin, A.V., Karnatovskaya, L.M., Chabueva, E.N., Pavlov, S.V., Kobrin, V.S., Starichenko, V.F., Kusov, S.Z., Knyazev, V.V., Boldyrev, V.V., Tolstikov, G.A. (2001) *Khim. Interesakh Ustoich. Razvit.* (2001) **9**, 625.

Etter, M.C., Frankenbach, G.M., Bernstein, J. (1989) *Tetrahedron Lett.* **30**, 3617.

Fernandez-Bertran, J., Reguera, E. (1998) *Solid State Ionics* **112**, 351.

Fernandez-Bertran, J., Reguera, E., Paneque, A., Yee-Madeira, H., Cordillo-Sol, A. (2002) *J. Fluorine Chem.* **113**, 93.

Forman, G.S., Tagmatarchis, N., Shinohara, H. (2002) *J. Am. Chem. Soc.* **124**, 178.

Gal'pern, E.G., Stankevich, I.V., Chistyakov, A.L., Chernozatonskii, L.A. (1997) *Chem. Phys. Lett.* **269**, 85.

Gillard, R.D., Pilbrow, M.F. (1974) *J. Chem. Soc., Dalton Trans.*, 2320.

Goiidin, V.V., Molchanov, V.V., Buyanov, R.A., Tolstikov, G.A., Lukashevich, A.I. (2003) *Russ. Pat.* 02196762.

Grohn, H., Paudert, R. (1963) *Z. Chem.* **3**, 89.

Gupta, M.K., Vanwert, A., Bogner, R.H. (2003) *J. Pharm. Sci.* **92**, 536.

Gutman, E.M., Bobovich, A.L. (1994) *Int. J. Mechanochem. Mechanic. Alloying* **1**, 153.

Hajipour, A.R., Arbabian, M., Ruoho, A.E. (2002) *J. Org. Chem.* **67**, 8622.

Hall, A.K., Harrowfield, J.M., Hart, R.J., McCormick, P.G. (1996) *Environ. Sci. Technol.* **30**, 3401.

Hasegawa, M., Kimata, M., Kobayashi, Sh.-I. (2002) *J. Appl. Polym. Sci.* **84**, 2011.

Heinicke, G. *Tribochemistry* (Carl Hanser Verlag, Munich, Germany, 1984).

Hihara, G., Satoh, M., Uchida, T., Ohtsuki, F., Miyamae, H. (2004) *Solid State Ionics* **172**, 221.

Ikekawa, A., Hayakawa, S. (1991a) *Funtai Kogaku Kaishi* **28**, 544.

Ikekawa, A., Hayakawa, S. (1991b) *Sib. Khim. Zh.*, 19.

Ikekawa, A., Ichikawa, J., Higuchi, Sh. (1990) *Zairyo Gijutsu* **8**, 171.

Ioffe, D.V., Ginzburg, I.M. (1983) *Khim. Prirodnykh Soedin.* 49.

Kalinovskaya, I.V., Karasev, V.E. (1998) *Zh. Neorg. Khim.* **43**, 1444.

Kalinovskaya, I.V., Karasev, V.E. (2003) *Zh. Neorg. Khim.* **48**, 1307.

Kashkovsky, V.I. (2003) *Katalis Neftekhimiya,* **11**, 78.

Kato, R. (2004) *Chem. Rev.* **104**, 5319.

Keshavarz, K.M., Knight, B., Srdanov, G., Wudl, F. (1995) *J. Am. Chem. Soc.* **117**, 11371.

Kolotilov, S.V., Addison, A.W., Trofimenko, S., Dougherty, W., Pavlishchuk, V.V. (2004) *Inorg. Chem. Commun.* **7**, 485.

Komatsu, K., Fujiwara, K., Murata, Ya., Braun, T. (1999) *J. Chem. Soc., Perkin Trans.* **1**, 2963.

Komatsu, K., Fujiwara, K., Tanaka, T., Murata, Y. (2000) *Carbon* **38**, 1529.

Komatsu, K., Wang, G.W., Murata, Ya., Tanaka, T., Fujiwara, K. (1998) *J. Org. Chem.* **63**, 9358.

Kondo, Sh.-I., Sasai, Ya., Hosaka, Sh., Ishikawa, T., Kuzuya, M. (2004) *J. Polym. Sci., Part A: Polym. Chem.* **42**, 4161.

Korolev, K.G., Golovanova, A.I., Mal'tseva, N.N., Lomovskii, O.I., Salenko, V.L., Boldyrev, V.V. (2003) *Khim. Interesakh Ustoich. Razvit.* **11**, 499.

Korolev, K.G., Lomovskii, O.I., Uvarov, N.F., Salenko, V.L. (2004) *Khim. Interesakh Ustoich. Razvit.* **12**, 339.

Krasnov, A.P., Mit', V.A., Afonicheva, O.V., Rashkovan, I.A., Kazakov, M.E. (2002) *Trenie Iznos* **23**, 397.

Krasnov, A.P., Rusanov, A.L., Bazhenova, V.B., Moroz'ko, A.N., Bulycheva, E.G.: Konarova, L.I., Doroshenko, Yu. E. (2003) *Plast. Massy*, No **6**, 23.

Li, J.C., Li, G.S., Wang, C., Zhang, Y.Q., Li, Y.L., Yang, L. H. (2002) *China J. Org. Chem.* **22**, 905.

Lomovskii, O.I., Korolev, K.G., Kwon, Y.S. (2003) *Proceedings — KORUS 2003, the Korea-Russia International Symposium on Science and Technology*, **1**, 7.

Lu, J., Li, Y., Bai, Y., Tian, M. (2004) *Heterocycles* **63**, 583.

Lukashevich, A.I., Molchanov, V.V., Goiidin, V.V., Buyanov, R.A., Tolstikov, G.A. (2002) *Khim. Interesakh Ustoich. Razvit.* **10**, 151.

Luty, T., Ordon, P., Eckhardt, C.J. (2002) *J. Chem. Phys.* **117**, 1775.

Ma, Ch., Chen, W.-Yo, Peng, Y.-H. (2004) *U.S. Pat.* 0092760.

Makhaev, V.D., Borisov, A.P., Alyoshin, V.V., Petrova, L.A. (1998) *Khim. Interesakh Ustoich. Razvit.* **6**, 211.

Masuda, S., Masame, K. (2001a) *Jpn. Pat.* 2001047026.

Masuda, S., Masame, K. (2001b) *Jpn. Pat.* 2001047028.

Mikhailenko, M.A., Shakhtshneider, T.P., Boldyrev, V.V. (2004a) *J. Mater. Sci.* **39**, 5435.

Mikhailenko, M.A., Shakhtshneider, T.P., Boldyrev, V.V. (2004b) *Khim. Interesakh Ustoich. Razvit.* **12**, 371.

Mit', V.A., Krasnov, A.P., Kudasheva, D.S., Afonicheva, O.V., Lioznov, B.S., Dubovik, I.I., Komarova, L.I. In: *Focus on Chemistry and Biochemistry.* Edited by Zaikov, G.E., Lobo, V.M.M., Garrotxena, N. (Nova Science Publishers, Hauppauge, NY, 2003, p. 93).

Mit', V.A., Krasnov, A.P., Kudasheva, D.S., Afonicheva, O.V., Lioznov, B.S., Dubovik, I.I., Komarova, L.I. In: *Chemical and Biochemical Kinetics: Mechanism of Reactions.* Edited by Zaikov, G.E. (Nova Science Publishers, Hauppauge, NY, 2004, p. 85).

Molchanov, V.V., Buyanov, R.A., Goiidin, V.V., Tkachyov, A.V., Lukashevich, A.I. (2002) *Kataliz v Promysht.*, No. 6, 4.

Molchanov, V.V., Goiidin, V.V., Buyanov, R.A., Tkachyov, A.V., Lukashevich, A.I. (2002) *Khim. Interesakh Ustoich. Razvit.* **10**, 175.

Mori, S., Kuriyama, O., Maki, Y., Tamai, Y. (1982) *Z. Anorg. Allg. Chem.* **492**, 201.

Moribe, K., Piyarom, S., Tozuka, Y., Yonemochi, E., Oguchi, T., Yamamoto, K. (2003) *STP Pharma Sci.* **13**, 381.

Moribe, K., Tsuchiya, M., Tozuka, Y., Yamaguchi, K., Oguchi, T., Yamamoto, K. (2004) *Chem. Pharm. Bull.* **52**, 524.

Mostafa, M.M., Abdel-Rhman, M.H. (2000) *Spectrochim. Acta* **56A**, 2341.

Murata, Ya., Kato, N., Fujiwara, K., Komatsu, K. (1999) *J. Org. Chem.* **64**, 3483.

Murthy, C.N., Geckler, K.E. (2001) *Chem. Commun. (Cambridge, UK)*, 1194.

Murthy, C.N., Geckler, K.E. (2002) *Fullerenes, Nanotubes, Carbon Nanostruct.* **10**, 91.

Nagata, N., Saito, F., Chang, Ch.-W. (2001) *Jpn. Pat.* 2001253969.

Nakai, Y., Yamamoto, K., Terada, K., Akimoto, K. (1984) *Chem. Pharm. Bull.* **32**, 685.

Nichols, P.J., Raston, C.L., Steed, J.W. (2001) *Chem. Commun. (Cambridge, UK)*, 1062.

Nomura, Yu., Nakai, S., Lee, B.-D, Hosomi, M. (2002) *Kagaku Kogaku Ronbunshu* **28**, 565.

Nuchter, M., Ondruschka, B., Trotzki, R. (2000) *J. Pract. Chem.* **342**, 720.

Oguchi, T., Kazama, K., Fukami, T., Yonemochi, E., Yamamoto, K. (2003) *Bull. Chem. Soc. Jpn.* **76**, 515.

Ohshita, T., Tsukamoto, A., Senna, M. (2004) *Physica Status Solidi A: Appl. Res.* **201**, 762.

Oprea, C.V., Popa, M. (1980) *Angew. Makromol. Chem.* **90**, 13.

Oprea, C.V., Simonescu, C. (1972) *Plaste Kautschuk* **19**, 897.

Palenic, G.F. (1969) *Inorg. Chem.* **8**, 2744.

Paneque, A., Fernandez-Bertran, J., Reguera, E., Yee-Madeira, H. (2001) *Transition Met. Chem.* **26**, 76.

Paneque, A., Fernandez-Bertran, J., Reguera, E., Yee-Madeira, H. (2003) *Synth. React. Inorg. Metal-Org. Chem.* **33**, 1405.

Paneque, A., Reguera, E., Fernandez-Bertran, J., Yee-Madeira, H. (2002) *J. Fluorine Chem.* **113**, 1.

Patchkovskii, S., Thiel, W. (1998) *J. Am. Chem. Soc.* **120**, 556.

Pauli, I.A., Poluboyarov, V.A. (2003) *Khim. Dizain*, 72.

Pecharsky, V.K., Balema, V.P., Wiench, J.W. (2003) *Int. Pat.* WO 066548.

Petrova, L.A., Borisov, A.P., Alyoshin, V.V., Makhaev, V.D. (2001) *Zh. Neorg. Khim.* **46**, 1655.

Petrova, L.A., Borisov, A.P., Makhaev, V.D. (2002) *Zh. Neorg. Khim.* **47**, 1987.

Petrova, L.A., Dudin, A.V., Makhaev, V.D., Zaitseva, I.G. (2004) *Zh. Neorg. Khim.* **49**, 645.

Pimentel, G.C. (1951) *J. Chem. Phys.* **19**, 446.

Piyarom, S., Yonemochi, E., Oguchi, T., Yamamoto, K. (1997) *J. Pharm. Pharmacol.* **49**, 384.

Pizzigallo, M.D.R., Napola, A., Spagnuolo, M., Ruggiero, P. (2004a) *Chemosphere* **55**, 1485.

Pizzigallo, M.D.R., Napola, A., Spagnuolo, M., Ruggiero, P. (2004b) *J. Mater. Sci.* **39**, 3455.

Pomogailo, A.D (2000) *Uspekhi Khim.* **69**, 60.

Qiu, W., Zhang, F., Endo, T., Hirotsu, T. (2004) *J. Appl. Polym. Sci.* **91**, 1703.

Rac, B., Mulas, G., Csongradi, A., Loki, K., Molnar, A. (2005) *Appl. Catal. A: Gen.* **282**, 255.

Ren, Zh., Cao, W., Ding, W., Shi, W. (2004) *Synth. Commun.* **34**, 4395.

Rowlands, S.A., Hall, A.K., McCormic, P.G., Street, R., Hart, R.J., Ebell, G.F., Donecker, P. (1994) *Nature* **367**, 223.

Saito, F., Chang, Ch.-W., Ouchi, A. (2002) *Jpn. Pat.* 2002030003.

Sangari, S.S., Kao, N., Bhattacharya, S.N., Silva, K. (2003) *61st Ann. Tech. Conf. Soc. Plast. Eng.* **2**, 1648.

Sasai, Ya., Yamaguchi, Yu., Kondo, Sh.-I., Kazuya, M. (2004) *Chem. Pharm. Bull.* **52**, 339.

Senna, M. (2002) *Khim. Interesakh Ustoich. Razvit.* **10**, 237.

Sereda, G.A., Zyk, N.V., Bulanov, M.N., Skadchenko, B.O., Volkov, V.P., Zefirov, N.S. (1996) *Izv. Akad. Nauk, Ser. Khim.* 2357.

Shandryuk, G.A., Kuptsov, S.A., Shatalova, A.M., Plate, N.A., Tal'roze, R.V. (2003) *Macromolecules* **36**, 3417.

Simonescu, C., Oprea, C.V., Nicoleanu, J. (1983) *Eur. Polym. J.* **19**, 525.

Strachan, A., van Duin, A.C.T, Chakraborty, D., Dasgupta, S., Goddard, W.A. (2003) *Phys. Rev. Lett.* **91**, 098301/1.

Sueishi, Y., Tobisako, H., Kotake, Ya. (2004) *J. Phys. Chem. B* **108**, 12623.

Suryanarayana, C. (2001) *Progr. Mater. Sci.* **46**,1.

Tanaka, T., Komatsu, K. (1999) *Synth. Commun.* **29**, 4397.

Tanaka, Y., Zhang, Q., Mizukami, K., Saito, F. (2003) *Bull. Chem. Soc. Jpn.* **76**, 1919.

Tanaka, Y., Zhang, Q., Saito, F. (2003a) *Ind. Eng. Chem. Res.* **42**, 5018.

Tanaka, Y., Zhang, Q., Saito, F. (2003b) *J. Phys. Chem. B* **107**, 11091.

Tanaka, Y., Zhang, Q., Saito, F. (2004) *J. Mater Sci.* **39**, 5497.

Taylor, L.J., Papadopoulos, D.G., Dunn, P.J., Bentham, A.C., Dawson, N.J., Mitchell, J.C., Snowden, M.J. (2004) *Org. Proc. Res. Dev.* **8**, 674.

Tipikin, D.S. (2002) *Zh. Fiz. Khim.* **76**, 518.

Tipikin, D.S., Lazarev, G.G., Lebedev, Ya.S. (1993) *Zh. Fiz. Khim.* **67**, 176.

Toda, F., Suzuki, T., Higa, S. (1998) *J. Chem. Soc., Perkin Trans.* **1**, 3521.

Todor, I.N., Orel, V.E., Mikhailenko, V.M., Danko, M.I., Dzyatkovskaya, N.N. (2002) *Eksp. Onkol.* **24**, 234.

Urano, M., Wada, Sh., Suzuki, H. (2003) *Chem. Commun. (Cambridge, UK)*, 1202.

Van Eldik, R., Klaerner, F.-G., Eds. *High Pressure Chemistry: Synthetic, Mechanistic, and Supercritical Applications* (Wiley-VCH Verlag GmbH & Co. KGaA, Weinheim, Germany, 2003).

Wada, Sh., Urano, M., Suzuki, H. (2002) *J. Org. Chem.* **67**, 8524.

Wang, G.W., Chen, Zh.X., Murata, Ya., Komatsu, K. (2005) *Tetrahedron* **61**, 4851.

Wang, G.W., Murata, Ya., Komatsu, K., Wan, T.S.M. (1996) *Chem. Commun. (Cambridge, UK)*, 2059.

Wang, G.W., Zhang, T.H., Hao, E.H., Jiao, L.J., Murata, Ya., Komatsu, K. (2003) *Tetrahedron* **59**, 55.

Watanabe, T., Ohno, I., Wakiyama, N., Kusai, A., Senna. M. (2002) *Int. J. Pharmaceuticals* **241**, 103.

Watanabe, T., Wakiyama, N., Kusai, A., Senna, M. (2004a) *Ann. Chim.* **29**, 53.

Watanabe, T., Wakiyama, N., Kusai, A., Senna, M. (2004b) *Powder Technol.* **141**, 227.

Webb, R.J., Dong, T.-Y., Pierpont, C.G., Boone, S.R., Chadha, R.K., Hendrickson, D.N. (1991) *J. Am. Chem. Soc.* **113**, 4806.

Yakusheva, L.D., Dubinskaya, A.M. (1984) *Biofizika* **29**, 365.

Yuan, Y., Wang, X., Xiao, P. (2004) *J. Eur. Ceramic Soc.* **24**, 2233.

Zaitsev, B.N., Dushkin, A.V., Boldyrev, V.V. (2001) *Izv. Akad. Nauk, Ser. Fiz.* **65**, 1292.

Zaitseva, I.G., Kuz'mina, N.P., Martynenko, L.I., Makhaev, V.D., Borisov, A.P. (1998) *Zh. Neorg. Khim.* **43**, 805.

Zhang, P., Pan, H., Liu, D., Guo, Zh.-X., Zhang, F., Zhu, D. (2003) *Synth. Commun.* **33**, 2469.

Zhang, Zh., Wang, X., Li, Zh. (2004) *Synth. Commun.* **34**, 1407.

Zhao, J., Feng, Y., Chen X. (2002) *Polym.-Plast. Technol. Eng.* **41**, 723.

Zhao, J., Feng, Y., Chen X. (2003) *J. Appl. Polym. Sci.* **89**, 811.

5 Mechanically Induced Phase Transition and Layer Arrangement

5.1 INTRODUCTION

Chapter 5 considers mechanically controlled phase transition, layer transition, order–disorder rebuilding, and swelling of layers or mixtures containing organic molecules (monomers or polymers). When surfaces are modified, they change adhesive, lubricating, viscous, and wetting properties as well as chemical affinity and biocompatibility. This modification allows for expanding the area of material applications toward microelectronics, smart surfaces, information storage, and medical devices.

5.2 LIQUID CRYSTALS

Liquid crystals are materials that exhibit characteristics of both liquids and crystalline solids. They possess quantitatively nonequivalent properties in different directions because their orientation is ordered. Some organic compounds display an organized structure between their solid and isotropic liquid states. Within this intermediate range, the mesophase, although physically behaving as a liquid, shows properties of an organized medium in the sense of the average orientation of molecules relative to a surface. For our consideration, it is important that warming of the mesophase up to the higher boundary of its existence does not change the orientation phase with respect to the surface (Nguyen et al. 2004).

Regarding physical behavior, liquid crystals form two main groups. The first group combines compounds with the ability to exist in the liquid crystalline state within a given temperature range. As noted, this state is intermediate between the molecular solids and the isotropic liquids. The compounds of this group are called *thermotropic liquid crystals*. The second group combines compounds with the ability to show the liquid crystalline state in a given concentration range. Such compounds are in intermediary forms between the solids and the dilute homogeneous solutions. They are called *liotropic liquid crystals*.

The course of preferential orientation of liquid crystalline molecules is characterized with the axial unitary vector and termed the *director*. The direction and degree of the liquid crystal orientation can be manipulated by applied external fields (magnetic, electrical); changes in temperature, pressure, or concentration; incorporation of aligning agents; chemistry, charge, and topography of contacting surface; and mechanical means (e.g., shear and stretching). This section considers structural phase transition initiated by external mechanical stress and concisely reviews applications of such phenomena.

5.2.1 Molecules of a Rodlike Shape

Compounds with a rodlike shape are sometimes named *calamatic* (or *calamitic*) molecules. They have partial orientation order that is typical for the liquid crystalline state. Depending on the ordering degree, different types of liquid crystalline states or mesophases can be distinguished. The least-ordered liquid crystalline phase is the nematic phase, in which the molecular axis has some preferential average direction. The direction can easily be changed on action of various external factors because liquid crystalline ensembles are anisotropic in the sense of viscoelastic, optical, electric, or magnetic properties. In turn, the direction change entails changes in optical, electrical, and other properties of the liquid crystal. Opportunities appear for manipulating these properties by relatively weak external effects. Conversely, registration of such effects is possible. Because of that, liquid crystals in the nematic phase have found diverse technological applications, for example, in liquid crystal displays used in calculator windows and computer monitors.

In the smectic phases, the molecules possess more order than in the nematic phase. Just as in the nematic phase, the molecules have their long molecular axis more or less parallel. In addition, the molecules are confined in layers. There are many types of smectic mesophases; two, smectic A and smectic C, are chosen as examples. In the smectic A phase, the direction is normal to the smectic layers. The individual molecules can be tilted with respect to the normal, but the average deviation from the normal angle is zero. In the smectic C phase, the direction does keep a definite average tilt angle with the normal one.

Shear induces structural changes of the liquid crystal layer. The changes are dependent on both the shear stress and the temperature. This leads to changes in the layer's viscosity. The molecular orientation depends on the product of the sliding velocity U and the layer thickness D, that is, UD. The molecular orientation is affected by solid surfaces; the behavior is known as *surface anchoring*. The effect of surface anchoring is stronger for lower UD (Nakano 2003). The temperature dependence of viscosity is sensitive to the change of the mean direction but not to the change of the precessional motion of the director (Negita and Uchino 2002).

Zheng et al. (2000) described transformations of cryogenic vesicles to threadlike micelles under shear for the mixture of an aromatic compound with a [(long-chain alkyl)trimethyl] ammonium chloride. The threadlike-to-vesicle transition is also possible. Shear and heating provoke the threadlike micellar structures to transform into spherical micelles (Lin et al. 2002). Namely, the formation of multilamellar vesicles took place when an aqueous solution of nonionic surfactant triethylene glycol monodecyl ether was put to successive shear flow turns (Nettesheim et al. 2004). The inherent nonequilibrium nature of such transition poses problems in technological applications of some liquid crystal formulations.

Water-soluble rodlike polymers acquire ordering alignment under the action of shear. The relevant example is 2–5% water solution of poly(2,2′-disulfonylbenzidine terephthalimide), shown in Scheme 5.1 (Funaki et al. 2004).

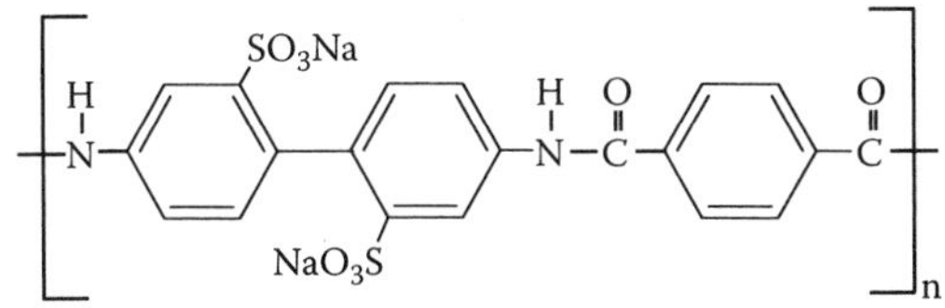

SCHEME 5.1

The liquid crystalline state in polymer solutions is caused by the highly rigid macromolecules. Ordering of the rigid macromolecules leads to an abrupt decrease in viscosity, which is important in respect to shear effects. Section 5.2.4 considers polymers containing liquid crystal moieties in more detail.

Once oriented in water solution, low molecular liquid crystals keep their orientation after removal of the solvent. Usually, shear aligns the orientation along the shearing direction. This point of view is commonly accepted (Schneider and Kneppe 1998). Therefore, works by Iverson et al. (1999, 2002) have conceptual importance and should be especially considered. The authors established ordering relative to the director not lengthwise, but at the right angle to the shearing direction for the liquid crystalline phase consisting of cationic perylene diimide with the ammonium terminal groups (Scheme 5.2).

This cationic diimide is soluble in water and displays unprecedented control of molecular orientation in solid films. The wet films were deposited on a glass plate and exposed to the action of mechanical shear. On removal of the water by evaporation under ambient conditions, the macroscopic alignment of the liquid crystal was transferred to the solid state. An anisotropically ordered solid film linearly polarized the light. The intense absorption of visible light polarized along the long axis of the molecule suggests that a significant number of molecules have a component of their electronic transition moment (which is parallel to the long axis of the molecule) projected normal to the shearing direction. Small-angle x-ray diffraction does indicate an increased order in the sheared solid film over the liquid nematic phase.

There is Friedel–Creagh–Kmetz regularity, which is known as the FCK rule (Creagh and Kmetz 1973; Friedel 1922): Liquid crystals align parallel to a surface of high energy (above 35 mJ/m^2) and perpendicular to a surface of low energy (less than the borderline value mentioned). Most surfaces, even polymers, have high surface energy, and generally liquid crystals of the rodlike type align parallel to the solid surface spontaneously.

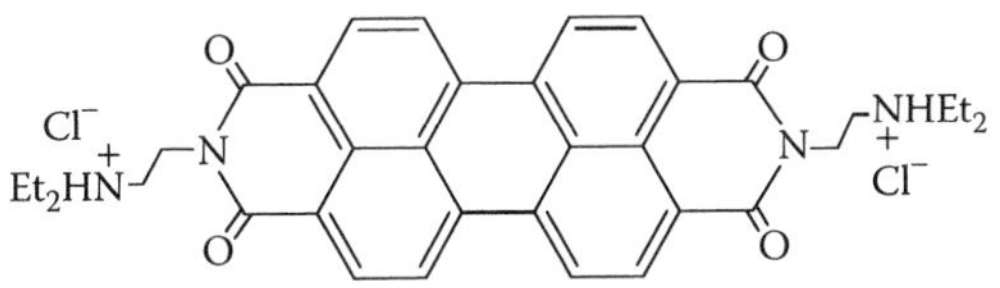

SCHEME 5.2

Surface treatment such as rubbing promotes a homogeneous orientation parallel to the surface. This results in improved lubrication. The enhanced lubrication effect is understandable in the sense of rolling or sliding. However, it is essential only for friction on a moderate load. At large load, liquid layers can penetrate each other, and the friction energy becomes significantly higher. Fortunately, microdevices preferentially work in non-large-load circumstances.

The essential function of lubricants is to separate two moving surfaces from each other, thereby reducing the friction energy. Importantly, smectic or nematic liquid crystals provide the possibility of perpendicular or twist disposition of their longitude axis in respect to the molecular layers, which slide past each other. In this case, the liquid crystal layers remain separated even at high vertical pressures. There are two ways to achieve the perpendicular or twist (so-called homeotropic) orientation: by application of an electric field perpendicular to the surface and by adsorbing surfactants on the solid surface.

The Kramer effect (electron exoemission; see Sections 3.2, 4.2, and 4.7) should also be mentioned. The liquid crystal molecule usually has donor and acceptor groups on opposite ends. Once captured, an exoelectron is preferentially localized on the fragment orbital of the acceptor moiety and endows it with the negative charge. Losing an electron, the surface acquires point-positive charge. Mutual attraction arises from the positively charged point of the surface and the negatively charged group of the liquid crystal molecule. In other words, the Kramer effect provokes the homeotropic orientation of the molecule with respect to the surface. Thus, in p-alkoxybiphenyl-p'-carbonitrile $RO-C_6H_4-C_6H_4-C\equiv N$, just the carbonitrile group will adhere to the surface. The long-chain octyloxy group will be immersed into a lubricating layer (Cognard 1990).

Fatty acids and their salts, lecithin, and other surfactants are alternatively used (Nishikawa et al. 1997). In this case, the molecular orientation of the liquid crystal near the solid surface is controlled in one direction by surface treatments (i.e., the surface-anchoring effects). If, however, the surfactant itself has a rodlike structure, the longer one (the liquid crystal or the surfactant) will be more perfectly ordered. An elegant experimental confirmation of this rule was obtained by Lammi et al. (2004). Moreover, the surfactant, because of its mobility, may act as a lubricant on the principal layer (Qian et al. 2004).

Nakano (2003) used infrared spectroscopy to investigate surface-anchoring effects on orientation of p-pentylbiphenyl-p'-carbonitrile (5-CB) in the absence of an external field (no electric potential, no mechanical action). Two kinds of pretreated surfaces were prepared: (1) glass plates coated with polyimide films and rubbed in one direction with a polyester cloth; and (2) glass plates on which cetyltrimethylammonium ammonium (CTAB) was chemically adsorbed. The polyimide-coated plates made the long axes of the 5-CB molecules parallel to the rubbing direction. The CTAB-coated plates made the 5-CB molecules perpendicular to the surfaces. In other words, two kinds of surfaces modified with 5-CB were prepared: parallel and perpendicular oriented. Clearly, such a difference in orientation affects friction conditions. Namely, at moderate velocity or load, the lower friction coefficient can be achieved with perpendicular orientation of liquid crystal lubricant.

The parallel orientation of a liquid crystal additive can also be achieved without pretreatment as a result of friction only. Desbat and coworkers (2003) showed that, for a rodlike liquid crystal deposited from chloroform solution on untreated Teflon film. After evaporation of the solvent and friction, the liquid crystalline layer (0.1- to 100-nm thick) presented good organization in the sliding direction.

5.2.2 Molecules of a Helixlike Shape

Helixlike nematic liquid crystals are termed *cholesterics*. Cholesteric molecules are oriented parallel to each other. However, there is additional twisting in the direction normal to their long axis. As a whole, the coiled structure is established as a spiral with a definite pitch. Thus, the axes of molecules of fatty acid esters of cholesterol are oriented on the rigid surfaces along the grooves of the surface microrelief. The hollows and bulges of the relief of the running-in pairs are reduced to the size of the cholesterics. After the optimal microrelief of the running-in surface is formed, wear ceases. When such relief is attained, the running-in surfaces are covered with a continuous film of liquid crystals. The parallel orientation of liquid crystal longitudinal axes in neighboring layers is caused by the guest–host effect. Because of such properties, liquid crystals like cholesterol reduce the friction coefficient of moving parts 5-fold and wear 20-fold or more (Vekteris and Murchaver 1995).

Solid surfaces exposed to the ambient atmosphere are covered with a layer of complex composition consisting mainly of water and the products of water/carbon dioxide interaction ($H_3O^+ + CO_3^-$). Such layer has surface tension of about 40 mJ/m^2 and is not displaced by most liquid crystals. This explains their orientation parallel to the solid surface (Cognard 1990). At low speed or low pressure, the surface layer shears in a plane of other (many) molecular layers distant from the surface. Actually, the liquid crystals do not interact directly with the surface, which remains covered with its atmospheric layer. They lie on it and layered asperities.

It is worth noting that the lubricating action of cholesteric liquid crystals is enhanced with time of exploitation. Liquid crystal mixtures deposited over a metal surface (steel, brass) spread and are well preserved for several months with no supplemental addition. The initial friction coefficient is diminished by a factor of 1.3 to 1.5 after 500 h of friction (Cognard 1990).

5.2.3 Molecules of a Disclike Shape

Disclike molecules can form liquid crystals in which such molecules are packed in stacks of the regular or irregular columnar type. In the regular columnar crystals, there is the long-range order in orientation of the discotic molecular planes. Such long-range order is absent in the irregular columnar crystals.

Pressure as a mechanical stress changes intercolumnar distance. Such effect was observed in the case of 2,3,6,7,10,11-hexakis(hexylthio)triphenylene (Maeda et al. 2003) (Scheme 5.3). Namely, inhomogeneous packing of the crystals with different intercolumnar distances is arranged at a pressure up to 300 MPa. While the core aromatic moiety remains unchanged, the flexible hydrocarbon side-chain tails are

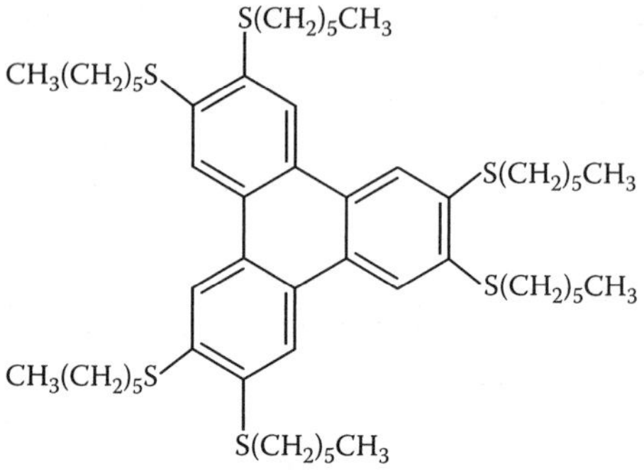

SCHEME 5.3

easily deformed under the stress. In turn, the distance between the discotic columns also changes.

5.2.4 Application Aspects of Liquid Crystal Lubricants

Of course, the relatively high cost of thermotropic liquid crystals predetermine their applications only for lubrication of friction parts of precise microdevices such as precise mechanical watches, micromotors, moving magnetic parts, and other such devices. The importance of these applications dictates the necessity to consider exploitation conditions.

5.2.4.1 Temperature Range

Liquid crystals turn into simple liquids at a certain temperature. They lose lubricating properties when the temperature is higher than the given maximum. At the same time, there are enough compounds with borderline points that are high enough for practical use. One can prepare a mixture of liquid crystals. After undercooling, these mixtures keep the liquid crystalline state for a long time before they freeze in the vitreous state. This extends their exploitation interval to low temperatures (although a high-temperature threshold is decreased). What is more important in the practical sense, pressure (loading) widens the temperature range for smectic and nematic ordering. The nematic order persists under pressure far above the isotropic transition point (Cognard 1990).

5.2.4.2 Load Range

At a very low load, the lubricating film is relatively thick, and the liquid crystals are imperfectly aligned. At a higher load, the film thickness decreases, and the shear rate increases. This leads to shear alignment of liquid crystalline layers. The rheological properties of the smectics/nematics are fully manifested only at the higher load (Fisher et al. 1988).

5.2.4.3 Anticorrosive Properties

Highly polar chemical substances are usually corrosive when they act as lubricants for metallic surfaces. Most liquid crystal molecules are polar. Liquid crystals are also dielectrics. Therefore, they eliminate the erosion processes, which cause electrochemical destruction of the surfaces of the mating pairs.

5.2.4.4 Coupled Action of Humidity and High Temperature (Tropical Conditions)

The majority of thermotropic liquid crystal compounds have the general formula of Y—Ar—X—Ar—Z. Here, X usually is N=N, N(O)=N, CH=N, CH_2—CH_2, CH=CH, C≡C, C(O)—NH; Y usually stands for Alk, AlkO, NH_2, and Z indicates Hal, CN, NO_2, C(O)OR groups. Although chemical structure (at the liquid crystalline state) does not play a role in lubricity, Schiff bases (—CH=N—), azoxy (—N(O)=N—) and carbonitrile (—CN) derivatives decompose in tropical conditions of high temperature and humidity. For instance, hydrolysis of the type Alk—Ar—Ar—CN → Alk—Ar—Ar—C(O)NH_2 was spectroscopically established (Cognard 1990). Ester derivatives [—C(O)OR] are more stable. They are also more available and cheaper than carbonitriles.

5.2.4.5 Mixing Liquid Crystal Additives with Base Oils

Friction and wear of aluminum–steel contacts were determined under variable conditions of applied load, sliding speed, and temperature in the presence of a lubricating base oil containing 1% liquid crystalline additives (Iglesias et al. 2004). In technique, aluminum alloy–steel contacts are especially difficult to lubricate, but the increasing number of applications of aluminum-based materials demands introduction of new additives (see, e.g., Mu et al. 2004). Iglesias and coauthors (2004) compared the tribological behavior of the base oil mixtures containing 1-dodecylammonium chloride, 4,4′-dibutylazobenzene, or cholesteryl linoleate. All of these formulations reduced the friction coefficient and the wear degree resulting from aluminum sliding against steel. Under increasing load, the ionic liquid crystal showed better lubricating behavior than the neutral ones. The ionic crystal provided stronger reduction of friction and wear at high sliding speed or temperature. In comparison with each of the additives used separately, a mixture of polar and nonpolar lubricants demonstrated better antifriction and antiwear properties. The effect is understandable from the consideration of adsorptive and surface-covering phenomena during lubrication.

Although thermotropic liquid crystals are expensive currently, liotropic ones have relatively low cost. For cost reasons, attention has turned to the liotropic group. As an industrial formulation, paraffin oils containing sorbitan monolaurate (SML) and ethoxylated sorbitan monolaurate (ESML) are promising. Of fundamental importance is the SML:ESML ratio. At the ratios of 7:3, 5:5, or 3:7, liquid crystal structures are formed very easily on conditions of lubrication (Wasilewski and Sulek 2003). The additive ratio to paraffin oil plays a decisive role in the formation of liquid crystalline structures in the surfactant phases. For instance, only a 0.5% concentration of 4-(4-hexylcyclohexyl)benzene isothiocyanate in paraffin oil provides the smallest friction coefficient on high load (Wazynska et al. 2004).

5.2.5 POLYMERIC LIQUID CRYSTALS

Liquid crystals contain mesogenic fragments, which are chemically inserted into the linear or comb-like macromolecular framework. The mesogenic groups of such macromolecules are easily oriented in the mesophase on external actions (mechanical, electric, magnetic). The presence of liquid crystal sequences imparts good operative properties for polymer liquid crystals. Disappearing low friction forces have been reported for solid surfaces bearing polymer brushes under appropriate (not swelling) solvent conditions and in the low contact pressure regime. Under good solvent conditions, long-range repulsive forces of osmotic origin act to keep the polymeric surfaces apart. Alongside entropic effects, this restricts mutual interpenetration of opposing polymer chains, thus maintaining a highly fluid layer at the interface. Under the high-pressure regime of some 100 MPa, compression of the brush becomes more significant, and the resistance to shear somewhat increases, but friction still remains low (Mueller et al. 2005; Yan et al. 2004).

Such peculiarities are technically promising. The ongoing process in several industries of replacement of metal parts and components by polymeric ones is slowing because polymeric surfaces undergo scratching and wear much more easily than metal surfaces. Although liquid lubricants can be used to lower friction and wear of moving metal parts without problems, this framework cannot be universally used for polymers. Polymer can take up a liquid and swell; the phenomenon is considered in Section 5.3.1. Dry friction of the entirely polymeric parts and those made from brushlike polymer, polyamides especially, is attracting increased attention (see, for example, Bermudez et al. 2005).

Mechanically induced reorientation of the polymeric liquid crystal direction was investigated (Merekalov and coauthors 2002). The acrylic polymer from Scheme 5.4 was used. On mechanical tension, the cholesteric spiral becomes uncoiled, and the polydomain-to-monodomain transition takes place.

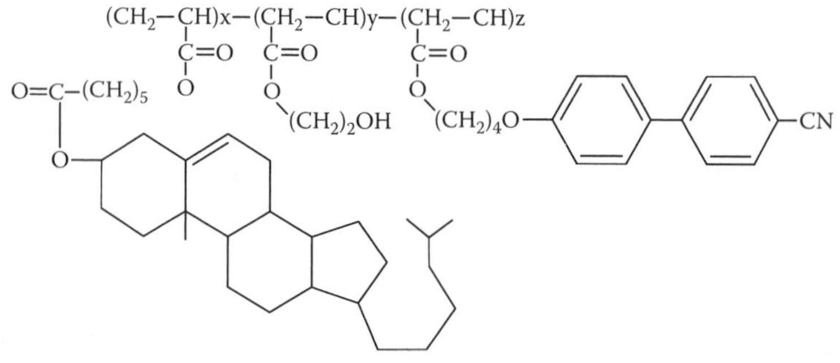

SCHEME 5.4

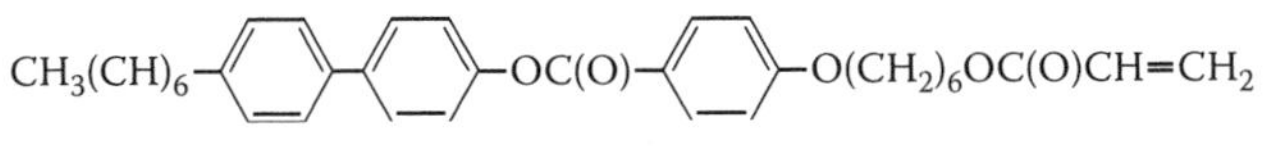

$$CH_3(CH)_6 - \text{⟨⟩} - \text{⟨⟩} - OC(O) - \text{⟨⟩} - O(CH_2)_6OC(O)CH=CH_2$$

SCHEME 5.5

Organic polymer thin films are widely used in displays as alignment layers to induce uniform, unidirectional orientation of liquid crystal molecules. This is critical to the optical and electrical performance of the corresponding devices.

Polymer liquid crystal alignment under mechanical actions is documented (Brostow et al. 1996 and references therein). If a side-chain liquid crystalline moiety is perpendicular to the polymeric backbone, shear flow induces parallel alignment, with the polymer and bent-down liquid crystalline fragments oriented in the flow direction (Pujolle-Robic and Noirez 2001). Shear also induces isotropic-to-nematic phase transition (Pujolle-Robic et al. 2002). The German patent claimed by Brinz (2002) gives one typical example of a liquid crystalline polymer. The polymer was manufactured by radical polymerization of the unsaturated acid ester depicted in Scheme 5.5. During friction, the polymer manufactured becomes oriented and shows strong antifriction properties.

Polymeric liquid crystals are also used in blends as they can serve as reinforcing components and as processing aids to decrease the viscosity of the blend. If these blends are formed by extrusion, the liquid crystalline components are finely distributed and aligned. This leads to improved mechanical properties. Thus, the extruded blend composed of polycarbonate (the matrix) plus 20% (by weight) of copolyester of poly(ethylene terephthalate) and *p*-hydroxybenzoic acid (liquid crystalline reinforcing component) demonstrates excellent tensile strength at shear rates up to 2000 sec^{-1} (Olszynski et al. 2002). The concentration of 20% of copolyester in the blend is optimal. A higher concentration of liquid crystal usually is not used because the cost increases rapidly, and phase inversion may take place (Kozlowski and La Mantia 1997). A low level of adhesion can weaken the reinforcing effect (Wei and Ho 1997), and compatibility of the blend can deteriorate (Brostow et al. 1996).

Importantly, the nature of the polymer layer can affect the orientation vector of the liquid crystal. Lee, Chae, et al. (2003) and Lee, Yoon, et al. (2003) correlated data on such orientation regarding molecular weight of the polystyrene used as an alignment layer. All the polystyrene samples had narrow polydispersity but a wide variety of molecular weights, from 2700 to 83,000. Rubbing of the polymer films orients the phenyl side groups preferentially perpendicular to the main chains. The main chains themselves are parallel to the rubbing direction (regardless of the molecular weight of the polymer sample). Rubbing generates grooves in the layer, and the polymer chains that pave the grooves meander. The larger the molecular weight is, the more meandering takes place because the polymer flexibility increases with the length of the backbone polymer chain. A solution of liquid crystal is then deposited on the oriented polymer layer. After evaporation of the solvent, orientation of the liquid crystal layers was studied by vibration spectroscopy and atomic force microscopy. The solution contained 4-pentyl-4′-cyanobiphenyl $\{p' - [CH_3(CH_2)_4]$ $-C_6H_4 - C_6H_4 - C\equiv N - p\}$ as the liquid crystal component. For polystryrenes with lower molecular weights (from 2700 to 5200), the liquid crystal molecules in contact with the rubbed polymer film surface are homogeneously induced to align nearly parallel to the rubbing direction. The rubbed films of the higher molecular weight polymer (9800 and higher) align the liquid crystal molecules in a nearly perpendicular manner in relation to the rubbing direction. The enhanced meandering of the longer polymer chain generates higher convexity of the groove-like structure. In the last case, cooperation between the liquid crystal and the meandering polymer groove leads to the vertical directional vector. The phenomenon should be taken into account when polymer supports for liquid crystal displays are needed.

Again, the unique rheology of liquid crystals makes them exceptional lubricants, especially in micromechanics. At this point, a patent by Mori (1995) should be mentioned. The patent claims the composition comprised of alkyl aryl nitriles (liquid crystal compounds) and perfluoroalkyl polyethers (oils). These compositions have high heat resistance, chemical stability, and low friction coefficients. In addition, they do not pollute the working environment.

5.3 POLYMERS

Incorporation of the problem of phase transition and allignment of non-liquid crystal polymers in the book seems to be useful for a unified consideration. A number of articles and books, such as *Mechanochemistry of Macromolecular Compounds* by Simonescu and Oprea (1970) or that by Baramboim (1971), reviewed the field. Section 5.3 updates coverage of this active area and provides references to more recent works. Naturally, the newly developed topics are emphasized.

5.3.1 SWELLING–DESWELLING

One new topic is a mechanochemical process for self-generation of rhythms and forms (see Boissonade 2003 and references therein). This is a general case when a mechanical event governs an "independent" chemical transformation. Specificity of the case consists of the reaction with some observable induction period. Three types of constituents can be considered: solvents (e.g., water), solutes (the reactants), and polymers (organized in a network, a gel). Swelling of the gel controls the transfer of chemicals inside and through the gel membrane. Swelling is a result of absorption of a liquid by a polymer. The process takes time. Many polymers are prone to time-dependent behavior in the swelling–deswelling process. This needs to be considered when designing such mechanochemical systems (Itano et al. 2005). After the induction period, interaction between the chemicals leads to product formation. If this results in volume contraction, the gel shrinks. Such swelling–deswelling phenomena naturally provide a coupling mechanism between the chemical processes operating within the gel and control the gel geometrical characteristics. For an appropriate initial size and sufficient swelling–deswelling amplitude, the process repeats indefinitely, so that the gel exhibits periodic changes of volume. Even a nonoscillating chemical reaction (starting after some induction time) can cause oscillatory instability,

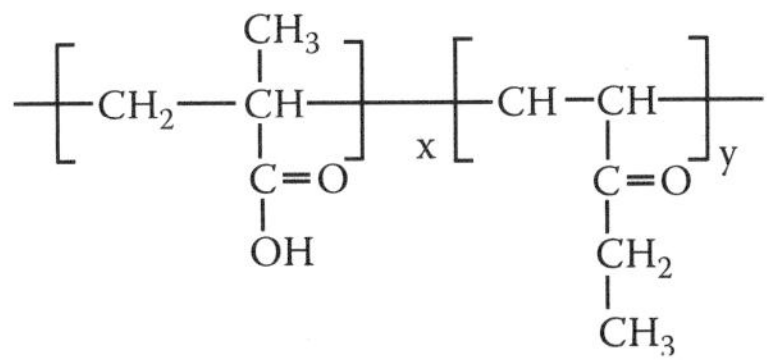

$$-[CH_2-\underset{\underset{OH}{\overset{\overset{CH_3}{|}}{C=O}}}{\overset{|}{CH}}-]_x [-CH-\underset{\underset{CH_3}{\overset{CH_2}{|}}}{\overset{\overset{}{}}{C=O}}CH-]_y-$$

SCHEME 5.6

leading to periodic changes of the gel's volume or shape. Because they are mechanically idle and chemically inert, polymeric gels are regularly used as supporting materials for the production of sustained reaction–diffusion patterns that are fed with fresh reactants by diffusion from the gel boundaries. Most rhythms and changes of shape in biological systems are governed by coupling of a chemical reaction with a mechanical event.

The polymeric gel itself also is a chemical compound. Its reactions can be connected with the swelling–deswelling transition. Neutralization and swelling of random methacrylic acid copolymer with ethyl acrylate is a pertinent example (Wang et al. 2005) (see Scheme 5.6 for the copolymer structure).

This polymer exists in compact latex particles that do not undergo swelling in pure water. However, swelling takes place effectively (although in a gradual manner) in water containing sodium hydroxide. The copolymer exists in a globular form because of hydrophobic association between ethyl acrylate blocks. Because the energy for extracting a proton from the carboxylic group within the globule is high, the carboxylic groups cannot freely dissociate in the water medium. In the nondissociated form, electrostatic repulsion of the polymer segments is absent, which keeps the polymer in a compact conformation. Sodium hydroxide is able to defeat the energetic barrier and to remove protons gradually from the carboxylic groups, endowing them with negative charges. As a result, a strong Coulomb potential arises around polyanions. Over the course of neutralization, the globule-to-coil conformational transition takes place. The new conformation is polysoaplike. Penetration of water molecules across the whole width of the polymer becomes possible. Swelling and dissolution come about, and the viscosity of carboxylic latex changes. The degree of such change depends on the depth of neutralization. The phenomenon described is of significant commercial importance because it provides a means to adjust the viscosity of polymeric latexes, for instance, in paint and adhesive formulations.

The perfluorinated ionomers composed of poly(tetrafluoroethylene) (PTFE) backbones with perfluorinated pendant chains terminated by sulfonic acid, such as Flemion, Nafion, and Aciplex, have been widely utilized in various industrial applications. Their structures can be expressed by the following general formula:

$$-(CF_2-CF_2)_x-(CF_2CF<)_y-[OCF_2CF(CF_3)]_m-O-(CF_2)_n-SO_3^-.$$

Specifically, they have attracted much attention as proton exchange membranes in polymer electrolyte fuel cells because the swollen membranes of these polymers have high proton conductivity. The proton conductivity markedly depends on the water uptake and mobility within the membrane. According to the result of molecular dynamic simulation (Urata et al. 2005), water mobility in a slightly wetted membrane is substantially restricted because of strong retention of water molecules with sulfonic groups. In contrast, in highly swollen membranes, even the water molecules located near sulfonic moieties are flexible enough and frequently exchange with relatively free water, which is located in the center of the membrane. This theoretical conclusion gives an approach for manufacturing the most active polymer electrolyte fuel cells.

Another important swelling–deswelling process is supercontraction of polymer fibers. In this process, an absorbed solvent induces large-scale transition from a glassy to a rubbery state, causing a macroscopic reduction in the length of the fiber and a concomitant swelling in diameter. Supercontraction in synthetic polymers is usually observed only at high temperatures or in solvents of extremely high dissolution activity.

However, one specific polymer fiber — spider silks — can supercontract at low ambient temperatures. In nature, supercontraction of spider silk fibers is induced by the before-dawn condensation of vaporous water from air. The water condensed on spider webs maintains their tension. Like other biomaterials, spider silk consists of repeated protein domains. Some of them are highly ordered; others are almost completely disordered. Spider silk is fibroin, that is, a protein polymer. This protein consists of repeating alanine- and glycine-rich regions. The alanine-rich regions are pleated sheets that form cross-links and provide strength and stiffness. Regarding the glycine-rich regions, they are less ordered and made from helices with threefold symmetry. This part lends elasticity.

According to van Beek et al. (2002), fibroin contains approximately 42% of glycine and 25% of alanine as the major amino acids. A full-circle (180°) turn occurs after each sequence of five glycine moieties. This provides the fibroin with spiral conformation (see also Asakura et al. 2005). The web is constructed from capture and dragline silk threads. The most elastic capture thread contains about 43 repeats and is able to extend by 200–400% in comparison to its original run. The dragline thread is constructed from shorter repeats (on average, 9 repeats) and is capable of extending by only 30% of its initial length.

During their passage through the narrowing tubes to the spinneret, the protein molecules align and partially crystallize at the expense of hydrogen bonding between them. Pleated sheets with highly ordered crystalline regions are formed. These sheets act as protein cross-links. They impart high tensile strength on the silk. On hydration, the hydrogen bonds stabilizing the sheets are broken. The chains collapse into mobile springs. So, the local phase transition takes place. Eles and Michal (2004) visualized such a transition by ^{13}C nuclear magnetic resonance spectra. [The authors studied webs from adult female *Nephila clavipes* spiders fed with (1-^{13}C-glycine)-enriched diet to obtain ^{13}C-labeled spider silk.]

The local phase transition leads to heterogeneity, which provides nucleation sites so that regions with low enough local stress are cooperatively hydrated and also collapse. As the fiber shrinks, the orientation distribution of the noncollapsed linkers widens. Because they contract on drying, the string regions lose their mobility because of replacement of water molecules with intra- and interchain hydrogen bonds. Such restoration of H bonds explains why spider silk undergoes hysteresis: when it is released from tension, it returns to shape.

The structural strength of the spider silk fibers attracts industrial interest. In 2002, Turner (of Nexia Biotechnologies, Inc., in Vaudreuil-Dorion, Quebec, Canada) outlined the *in vitro* microinjection of a genetically engineered segment of the spider DNA (deoxyribonucleic acid) into fertilized goat eggs. The female goats were born, grew up, and began to produce milk containing the desired protein. The crystallinity level of such milk-originated silk fibers can be controlled by doping agents. Applications of these fibers are expected to be in (1) the medical market for microsutures in ophthalmic and neurological surgeries (about 100 goat females are needed for silk production); (2) fishing lines (2000–4000 heads); and (3) lightweight bullet-proof vests (345,000 vests for the U.S. Army are needed, which means 5000–8000 goats are needed). Obviously, practical applications of spider silk fibers are not too distant.

5.3.2 MECHANICALLY DEVELOPED PREORIENTATION

5.3.2.1 Alignment on Grinding

After preparation, the biradical 5,5′-bis(1,3,2,4-dithiadiazolyl) does not display any ferromagnetic character. Grinding increases its magnetization as a function of treatment time. The room temperature effective magnetic moment increases 2.5 times on grinding for 12 h (Antorrena et al. 2002). The structure of the biradical monomer is depicted in Scheme 5.7.

The grinding likely provides the energy needed to overcome the activation barrier of the transition from diamagnetic to the more thermodynamically stable paramagnetic phase. The powder x-ray diffraction data for the diamagnetic nonground and the paramagnetic ground material are consistent with the second-order phase transition between the two phases. The phase change leads to structural transition from a black fluffy powder to black graphitelike shiny plates. Importantly, no changes in the Raman, infrared, and mass spectra are observable. The authors (Antorrena et al. 2002) assumed that polymerlike association exists between the biradical monomers before grinding. Grinding destroys such association. Quantum chemical calculations showed that there is a significant charge distribution between the nitrogen and oxygen atoms of the outer bond within the biradical framework. Namely, the sulfur atom is positively charged (+084), whereas the nitrogen atom is charged negatively (−0.87) (see Scheme 5.7). This charge distribution provides electrostatic attractions between

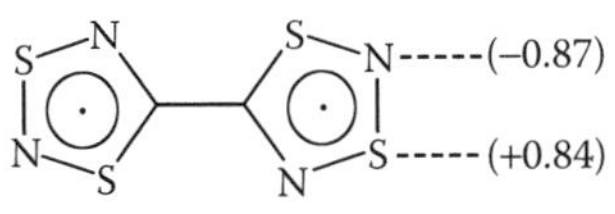

SCHEME 5.7

biradical species in the solid sample on the eve of grinding. It is grinding that destroys the ploymerlike chain and enhances ferromagnetic coupling of unpaired electrons in the sample.

5.3.2.2 Alignment on Brushing

One of the most important applications of polymers is their use as supporting layers for liquid crystal alignment. This section considers the other side of this problem, i.e., alignment of the polymeric support trials. Unidirectional molecular alignment is fundamental to operation of liquid crystal displays. At the interface with a solid, liquid crystal molecules are anchored and aligned. The interface commands the directional orientation of liquid crystals. For this purpose, the surface itself is preliminary oriented. Polymers with long-chain cores are appropriate for the preparation of such interface layers. Over several decades, a variety of alignment-promoting layers have been used on the inside surfaces of liquid crystal displays. Initially, thin layers of long-chain polymers, such as poly(vinyl alcohol), and organic molecules, such as organyl silanes, were applied to surfaces. These materials were then replaced by polyimides to give improved performance. At present, polyimides are mostly used because of their advantageous properties, such as excellent optical transparency, adhesion, heat resistance, dimensional stability, and insulation ability. A rubbing process using a rayon velvet fabric (a cellulose fabric) is the only technique adopted in the liquid crystal industry to treat such film surfaces for the mass production of flat-panel displays. Rubbing generates microgroove lines in the polyimide film. The line direction coincides with that of the polyimide backbone axis. In practice, the polyimide dilute solution in organic solvent (e.g., in N-methyl-2-pyrrolidinone) is deposited on a support slide. The samples were cured initially by soft backing, followed by thermal treatment. The cured samples were rubbed using a machine with a flat plate housing the sample. The sample was passed under a velvet-coated drum at a constant, predefined speed (for details, see Macdonald et al. 2003). Such oriented polymer film is covered with a liquid crystal to make an ultrafine film.

When a thermotropic liquid crystalline polymer is deposited on a rubbed polymer support, thermal treatment is recommended to align the director perfectly along the rubbed trail (Kinder et al. 2004).

The parallel alignment of the liquid crystal director and the support trail usually coincide. Nevertheless, there are cases when the director and the trail are mutually perpendicular. Chae and colleagues (2002) found such mutual orientation between a rodlike liquid crystal and a well-defined brush polymer composed of aromatic-aliphatic bristles set into fully rodlike polymer backbone. In this case, 4′-pentylbiphenyl-4-carbonitrile (5-CB), a liquid crystal, and poly[p-phenylene 3,6-bis(4-butoxyphenyloxo)pyromellitimide], a brush-polymer support, were used. The abnormal liquid crystal alignment is attributed to the strength of the anisotropic intermolecular interactions of the liquid crystal molecules with the short bristles attached perpendicular to the polymer chain. The crystal–bristle interactions override the interactions with the main polymer chains and with the microgroove lines created along the rubbing direction. Scheme 5.8 shows the described molecules and the mutual orientation of their axes.

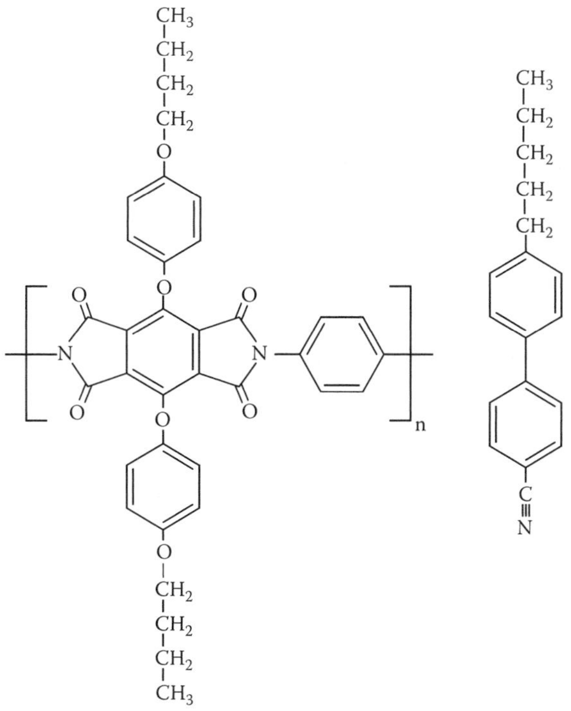

SCHEME 5.8

5.3.2.3 Alignment on Friction

Sometimes, friction initiates phase transitions of polymeric films that are responsible for their lubricating properties. PTFE is an example. Structurally, the polymer is a repeating chain of substituted ethylene with four fluorine atoms on each ethylene unit, namely $-(CF_2-CF_2)_n-$. From sliding contacts, the polymer acquires the ability to lubricate, showing an outstandingly low coefficient of friction (less than 0.1). The antifriction property is attributed to the smooth molecular profile of the polymer chains, which turn to be oriented in a manner that facilitates easy sliding and slipping. For polymers of the polyethylene family, high pressure initiates the formation of extended-chain modifications with enhanced mobility (Rein et al. 2004). On warming, PTFE transforms into rod-shape macromolecular particles that slip along each other, similar to lamellar structures. Chemical inertness makes PTFE useful in cryogenic to moderate operating temperatures and in a variety of atmospheres and environments. Operating temperatures are limited to about 260°C because of decomposition of the polymer. PTFE finds many uses in lubrication at ambient temperature. These applications include fasteners, thread elements, and so on. PTFE also serves as an additive in lubricating greases and oils (Mariani 2003).

Of course, the rodlike alignment is not a single effect of lubrication. In particular, PTFE does accept electrons — electrons that leave the metallic surfaces during their mutual sliding. In principle, such electron attachment can initiate (and initiates, indeed) degradation of PTFE. However, this process is delayed because a fraction of the negative charge remains in the PTFE (Wasem et al. 2003).

PTFE is an excellent filler for polymers used as self-lubricating, maintenance-free bearing materials for the manipulation and positioning of very-large-scale and heavy-loaded constructions, especially in offshore applications. For instance, friction of polyethylene terephthalate (PETP) against steel is characterized by a steady-state coefficient of 0.28 at a contact pressure of 55 N/mm². At the same conditions, samples of PETP filled with PTFE showed a coefficient of 0.09 (De Baets et al. 2002).

5.3.2.4 Alignment on Crystallization

Some unexpected results were obtained by Breiby et al. (2005) during friction of PTFE deposited on a glass plate. The initial cylindrical uniaxial structure of PTFE was changed to a wormlike one after the friction. Namely, the crystallites in the deposited film became highly biaxially oriented, not only *in* but also *about* the chain axis direction. Reasons and possible technical application of this phenomenon remain unclear.

One practically important application relates to mechanically provoked crystallization of thermoplastic polymers. In this case, crystallization means unified alignment of the polymer molecules. Crystallization of many thermoplastic polymers is necessary to make them workable in further processing steps toward preparation of plastic end products. This process is essential when the polymers are amorphous and obtained by copolymerization or blending. Such plastics become sticky at relatively low temperatures. Without crystallization, it is difficult to handle these polymer particles in blow molding, extrusion, and blending. The problems are brought about because of agglomeration of the polymeric masses during the operation. This usually results in the polymers sticking to operating equipment.

Thermoplastic polymers usually crystallize during slow heating. An intermediary sticky phase occurs at temperatures somewhat lower than the crystallization temperature of the given polymer. The polymer particles are slowly heated. The polymer chains are ordered within the crystal lattice. The ordering is accompanied by an exothermal effect. It is important to approach the exothermic step slowly to avoid a temperature jump. Even at slow heating, some part of the thermoplastics remains sticky and agglomerates to form large, untreatable masses. This can damage the crystallization equipment. The polymer properties (intrinsic viscosity, melting point, particle size) also deteriorate. With slow heating of the polymers, long-term elevated temperatures cause destruction and yellowing of the final product.

To circumvent all of these obstacles, a method is claimed to provide an effective, nonagglomerating crystallization (Moore 2002). In the method, mechanical deformation of thermoplastic polymer particles is performed in a cylindrical crystallizer equipped with fluidizing blades moving around, and in proximity to, the inner wall of the apparatus. When the blades move, mechanical friction takes place between the polymer particles, the apparatus wall, and the fluidization blades. This generates both heat and mechanical stress. The polymer molecules align into crystal lattices. For a period of time between 20 and 200 min, the mass temperature exceeds the crystallization temperature by 10–50°C. The crystal content achieved by the method is up to 80%. The product does not have the yellow color and is absolutely free of agglomerated granules, powder, or thin films. No polymer is sticking to the crystallizer at completion of the operation. The product is well prepared for further technological treatment.

5.3.2.5 Alignment on Stretching

Preorientation is aimed at forcing segments of polymer chains into a uniaxial direction, which improves tensile strength and yield stress as well as toughness. Such properties were improved by preorientation of Nodax film as a result of stretching (Hassan et al. 2004). Nodax is a random copolymer from Procter & Gamble Company (West Chester, OH). The copolymer consists of 87–89 mol% 3-hydroxybutyrate and 11–13 mol% 3-hydroxyhexanoate. Films were prepared by press molding of the copolymer into Teflon-coated aluminum molds at 130°C. Preorientation of the films was performed by a combination of heating at 130°C for 10 min, annealing for 5 min, and uniaxial stretching. The stretched sample was examined by differential scanning calorimetry, x-ray diffraction, and birefringence. At high draw ratios, the majority of the copolymer chains were unfolded and extended in the stretching direction. At the same time, some of the chains were perpendicular to the film surface. Thus, drawing leads to the formation of a new periodic structure in which the chain axis is preferentially parallel to the draw direction. This increases crystallinity and improves the material technological parameters if they are measured in the draw direction. At the same time, the copolymer remains weak perpendicularly. Hassan et al. (2004) suggested that a biaxial stretching procedure applied to Nodax film can provide more useful material that will be extremely strong in all directions. This definitely will move Nodax forward in industrial end-use applications.

5.4 PRESSURE-INDUCED PHASE TRANSITION

The application of high pressure to solids is a powerful way of changing their physicochemical properties. In this sense, small organic molecules attract considerable attention because they exhibit photoconductivity and electroluminescence. These properties make them promising materials for optoelectronic devices. In particular, high electron mobility has been found for anthracene in the crystal state (Karl and Marktanner 2001), making this substance a good candidate for applications. The strong anisotropy of conductivity and the dielectric function in the crystals is closely related to the specific manner of molecular packing. Anthracene is a highly compressible solid. Pressure alters the intermolecular interactions in a controlled way. By some reduction of the unit-cell volume, the overlap of the molecular orbitals is enhanced. This leads to stronger intermolecular interaction and interlayer packing in the crystal. The change was established by x-ray diffraction of the anthracene samples pressurized up to 22 GPa (Oehzelt et al. 2003). Related phenomena (leading to mechanochromism) were considered in Chapter 2. In the case of the anthracene crystals, pressure indeed induces red shifts in fluorescent spectra (Dreger et al. 2003). The pressure effects on interaction between the anthracene molecules and on their electronic and optical properties were evidenced by density functional calculations (Hummer et al. 2003).

An analogous situation pertains to pentacene, another representative of polycyclic aromatic hydrocarbons. Pentacene contains five benzene rings fused linearly. For its polymorph grown from the vapor phase, electron mobility is observed (Siegrist et al. 2001). Actually, the pentacene molecular crystals exist in two polymorphs characterized by a small difference in the minimal potential energies (not more than 1 kJ/mol). Both polymorphs belong to the same space group, but differ in the crystal lattice parameters (Della Valle et al. 2003). At normal pressure, the relative differences in the specific volumes of the two polymorphs are rather small, amounting to less than 1% (Farina et al. 2003). By applying a moderate pressure of only 0.2 GPa, structural phase transition of one form to another takes place (Farina et al. 2003).

Pressure-induced structural changes in crystalline oligo(para-phenylenes) were also reported (Heimel et al. 2003). The molecules studied were diphenyl, *p*-terphenyl, *p*-quaterphenyl, *p*-quinquephenyl, and *p*-sexiphenyl. All are technologically applicable as solids, mainly in optoelectronic devices. Their activity can be tuned by applying pressure. This is a "clean" way to change the degree of intermolecular interaction. At pressures up to 6 GPa, the molecular planes tilt more toward a bricklike alignment within one layer's herringbone pattern.

The application of pressure plays a significant role in changing the electronic states of low-dimensional molecular conductors, as demonstrated in the pressure-induced superconductivity of the tetramethyltetraselenafulvalene cation-radical salt with hexafluorophosphate anion and its family (Kato 2004). The applied pressure enhanced the intercolumn interactions, thus providing the dimensionality of the electronic structure to suppress the Peierls instability. Two representative examples of the ferromagnetic-to-antiferromagnetic phase transition in solid states are depicted on Scheme 5.9. In the case of the left example in Scheme 5.9 [4-(4-chlorobenzylideneamino)-2,2,6,6-tetramethylpiperidin-1-oxyl], pressure effect (0.5–07 GPa) leads to changes in the intermolecular chain contacts between the nitroxide oxygen of one molecule and methyl hydrogen atoms of another molecule. As a result, a loss of ferromagnetism is observed (Mito et al. 2001). In the case of the right example in Scheme 5.9 [iron bis(pyrimidine) dichloride], intramolecular orbital interactions are changed on 0.7-GPa pressure. The antiferromagnetic transition is a consequence of a stronger orbital overlap giving rise to an enhanced electron superexchange (Wolter et al. 2003).

The effect of pressure on amino acids is important for geo- and cosmochemistry. Using crystalline γ-glycine, Boldyreva et al. (2004) obtained a new high-pressure polymorph. The authors revealed that γ-glycine underwent a first-order phase transition accompanied by an abrupt change in the unit cell volume. The phase transition

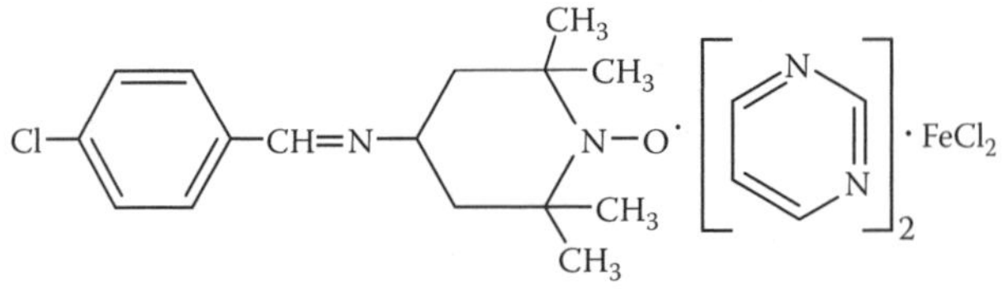

SCHEME 5.9

commenced at a pressure of 2.74 GPa and was not completed even at 7.85 GPa. The structure of the high-pressure phase is described in space group *Pn*. The glycine zwitterions in the structure are linked by hydrogen bonds to form layers, which are combined in pairs. The question of driving force for this phase transition remains open.

Solid triglycine sulfate ($HOOC—CH_2—NH_2)_3 \times H_2SO_4$ changes its dielectric constant and Raman spectrum with pressure (at room temperature). The ferroelectric and paraelectric phases coexist. After a 2.5-GPa pressure was reached, a new, so-called high-pressure phase, appeared. The lattice parameters of this phase were distinctly different from the starting ones. Structural phase transition occurred at around 2.5 GPa. This transition is conditioned by the order–disorder conversion in the solid (Suzuki et al. 2003).

Dynamic reorientation of the self-assembled monolayers was proposed as a measure of the monolayer elasticity and viscoelasticity under conditions of fast, large-amplitude uniaxial compression (Lagutchev et al. 2005; Patterson and Dlott 2005). That has been demonstrated for monolayers of octyl, pentadecyl, or benzyl mercaptane on gold or silver substrates.

5.5 CONCLUSION

In research into tribology, the possibility of using liquid crystals as lubricants has been investigated. They displayed excellent lubrication properties. A friction coefficient can be controlled by applying electric fields across liquid crystal lubricating films, such as in the control of optical characteristics in liquid crystal displays (Nakano 2002). According to the theme of this book, control of lubrication condition by thermotropic liquid crystals was scrutinized. The trend is also to use liotropic liquid crystals as additives to lubricating oils. These crystals are cheaper than thermotropic ones. Waste products in manufacture of thermotropic liquid crystals would be interesting to probe regarding lubrication.

Mechanically induced liquid crystal alignment is crucial for operation of diverse displays. Mechanical friction initiates crystallization of thermoplastics, helping to avoid many obstacles usually accompanying the polymer preparation for further technological treatments.

Swelling–deswelling processes in polymers promise many useful applications. Diffusion of chemicals into the gel membranes and their reactive interaction within the gel show definite potential. Changes in the gel space dimensions at the expense of swelling are also promising for practical applications. Coupling of chemical reactions with mechanical effects is a way to adopt biotransformations to human benefits. Such coupling is also important for adjustment of polymer-latex viscosity, for instance, in paint and adhesive formulations. Stretching of polymer films represents a way to improve their mechanical properties.

Using small molecules as electron carriers, optoelectronic devices can be tuned by applying pressure. Pressure also induces superconductivity of some radical ion salts. Dynamic reorientation of the self-assembled monolayers under pressure is an indicator of their elasticity, which is important for practical purposes.

To sum up, Chapter 5 demonstrates applicability of organic mechanochemical transitions to technical innovations.

REFERENCES

Antorrena, G., Brownridge, S., Cameron, T.S., Palacio, F., Parsons, S., Passmore, J., Thompson, L.K., Zarlaida, F. (2002) *Can. J. Chem.* **80**, 1568.

Asakura, T., Yang, M., Kawase, T., Nakazawa, Ya. (2005) *Macromolecules* **38**, 3356.

Baramboim, N.K. *Mechanochemistry of Macromolecular Compounds/Mekhanokhimiya Vysokomolekulyarnykh Soedinenii* (Khimiya, Moscow, 1971).

Bermudez, M.D., Brostow, W., Carrion-Vilches, F.J., Cervantes, J.J., Pietkiewich, D. (2005) *Polymer* **46**, 347.

Boissonade, J. (2003) *Phys. Rev. Lett.* **90**, 188302.

Boldyreva, E.V., Ivashevskaya, S.N., Sowa, H., Ashbahs, H., Weber, H.-P. (2004) *Dokl. Akad. Nauk* **396**, 358.

Breiby, D.W., Solling, Th.I., Bunk, O., Nyberg, R.B., Norrman, K., Nielsen, M.M. (2005) *Macromolecules* **38**, 2383.

Brinz, Th. (2002) *Germ. Pat.* 10102238.

Brostow, W., Sterzynski, T., Triouleyre, S. (1996) *Polymer* **37**,1561.

Chae, B., Kim, S.B., Lee, S.W., Kim, S.I., Choi, W., Lee, B., Ree, M., Lee, K.H., Jung, J.Ch. (2002) *Macromolecules* **35**, 10119.

Cognard, J. (1990) *ACS Symp. Ser.* **441** *(Tribol. Liq.-Cryst. State)*, 1.

Creagh, L.T., Kmetz, A.R. (1973) *Mol. Cryst. Liq. Cryst.* **24**, 59.

De Baets, P., Ost, W., Samyn, P., Schoukens, G., Van Parys, F. (2002) *Synth. Lubr.* **19**, 109.

Della Valle, R.G., Venuti, E., Brillante, A., Girlando, A. (2003) *J. Chem. Phys.* **118**, 807.

Desbat, B., Brunet, M., Nguyen, H.T. (2003) *Soft Materials* **1**, 75.

Dreger, Z.A., Lucas, H., Gupta, Y.M. (2003) *J. Phys. Chem. B* **107**, 9268.

Eles, Ph.T., Michal, C.A. (2004) *Macromolecules* **37**, 1342.

Farina, L., Brillante, A., Della Valle, R.G., Venuti, E., Amboage, M., Syassen, K. (2003) *Chem. Phys. Lett.* **375**, 490.

Fisher, T.E., Bhattacharya, S., Salher, R., Lauer, J.L., Ahn, Y.-J. (1988) *Tribol. Trans.* **31**, 442.

Friedel, G. (1922) *Ann. Phys.* **18**, 206.

Funaki, T., Kaneko, T., Yamaoka, K., Ohsedo, Y., Gong, J.P., Yoshihito, O., Shibasaki, Y., Ueda, M. (2004) *Langmuir* **20**, 6518.

Hassan, M.K., Abdel-Latif, S.A., El-Roudi, O.M., Sharaf, M.A., Noda, I., Mark, J.E. (2004) *J. Appl. Polym. Sci.* **94**, 2257.

Heimel, G., Pushnig, P., Oehzelt, M., Hummer, K., Koppelhuber-Bitschnau, B., Porsch, F., Ambrosch-Draxl, C., Resel, R. (2003) *Materials Research Society Symposium Proceedings, 771 (Organic and Polymeric Materials and Devices)*, 249.

Hummer, K., Puschnig, P., Ambrosch-Draxl, C. (2003) *Phys. Rev. B: Condens Matter Mater. Phys.* **67**, 184105/1.

Iglesias, P., Bermudez, M.D., Carrion, F.J., Martinez-Nicolas, G. (2004) *Wear* **256**, 386.

Itano, K., Choi, J., Rubner, M.F. (2005) *Macromolecules* **38**, 3450.

Iverson, I.K., Casey, S.M., Seo, W., Tam-Chang, S.-W., Pindzola, B.A. (2002) *Langmuir* **18**, 3510.

Iverson, I.K., Tam-Chang, S.-W. (1999) *J. Am. Chem. Soc.* **121**, 5801.

Karl, N., Marktanner, J. (2001) *Mol. Cryst. Liq. Cryst. Sci. Technol. Sect. A* **355**, 149.

Kato, R. (2004) *Chem. Rev.* **104**, 5319.

Kinder, L., Kanicki, J., Petroff, P. (2004) Synth. Met. **146**, 181.

Kozlowski, M., La Mantia, F.P. (1997) *J. Appl. Polym. Sci.* **66**, 969.

Lagutchev, A.S., Patterson, J.E., Huang, W., Dlott, D.D. (2005) *J. Phys. Chem. B* **109**, 5033.

Lammi, R.K., Fritz, K.P., Scholes, G.D., Barbara, P.F. (2004) *J. Phys. Chem. B* **108**, 4593.

Lee, S.W., Chae, B., Kim, H.Ch., Lee, B., Choi,W., Kim, S.B., Chang, T., Ree, M. (2003) *Langmuir* **19**, 8735.

Lee, S.W., Yoon, J., Kim, H.Ch., Lee, B., Chang, T., Ree, M. (2003) *Macromolecules* **36**, 9905.

Lin, Zh., Mateo, A., Zheng, Y., Kesselman, E., Pancallo, E., Hart, D.J., Talmon, Y., Davis, H.T., Scriven, L.E., Zakin, J.L. (2002) *Rheol. Acta* **41**, 483.

Macdonald, B.F., Zheng, W., Cole, R.J. (2003) *J. Appl. Phys.* **93**, 4442.

Maeda, Y., Rao, D.S.Sh., Prasad, S.K., Chandrasekhar, S., Kumar, S. (2003) *Mol. Cryst. Liq. Cryst.* **397**, 429.

Mariani, G. (2003) *Chem. Ind. (Dekker)* **90** *(Lubr. Add.)*, 171.

Merekalov, A.S., Otmakhova, O.A., Sycheva, T.I., Tal'roze, R.V. (2002) *Vysokomol. Soedin., Ser. A Ser. B* **44**, 1992.

Mito, M., Kawae, T., Hitaka, M., Takeda, K., Ishida, T., Nogami, T. (2001) *Chem. Phys. Lett.* **333**, 69.

Moore, W.P. (2002) *U.S. Pat.* 6479625.

Mori, M. (1995) *Jpn. Pat.* 0782582.

Mu, Z., Liu, W., Zhang, Sh., Zhou, F. (2004) *Chem. Lett.* **33**, 524.

Mueller, M.T., Yan, X.P., Lee, S.W., Perry, S.S., Spencer, N.D. (2005) *Macromolecules* **38**, 5706.

Nakano, K. (2002) *Ekisho* **6**, 137.

Nakano, K. (2003) *Lubr. Sci.* **15**, 233.

Negita, K., Uchino, S. (2002) *Mol. Cryst. Liq. Cryst. Sci. Technol., Sec. A: Mol. Cryst. Liq. Cryst.* **378**, 103.

Nettesheim, F., Olsson, U., Lindner, P., Richtering, W. (2004) *J. Phys. Chem. B* **108**, 6328.

Nguyen, Th.-Q., Martel, R., Avouris, Ph., Bushey, M.L., Brus, L., Nuckolls, C. (2004) *J. Am. Chem. Soc.* **126**, 5234.

Nishikawa, M., Miyamoto, T., Kawamura, Sh., Yasuda, K., Mutsuga, Ya., Matsuki, Ya. (1997) *U.S. Pat.* 5700860.

Oehzelt, M., Heimel, G., Resel, R., Puschnig, P., Hummer, K., Ambrosch-Draxl, C., Takemura, K., Nakayama, A. (2003) *Materials Research Society Proceedings 771 (Organic and Polymeric Materials and Devices)*, 219.

Olszynski, P., Kozlowski, M., Kozlowska, A. (2002) *Mater. Res. Innovation* **6**, 1.

Patterson, J.E., Dlott, D.D. (2005) *J. Phys. Chem. B* **109**, 5045.

Pujolle-Robic, C., Noirez, L. (2001) *Nature* **409**, 167.

Pujolle-Robic, C., Olmsted, P.D., Noirez, L. (2002) *Europhys. Lett.* **59**, 364.

Qian, L., Charlot, M., Perez, E., Luengo, G., Potter, A., Cazeneuve, C. (2004) *J. Phys. Chem. B* **108**, 18608.

Rein, D.M., Shavit, L., Khalfin, R.L., Cohen, Ya., Terry, A., Rasogy, S. (2004) *J. Polym. Sci., Part B: Polym. Phys.* **42**, 53.

Schneider, F., Kneppe, H. In: *Handbook of Liquid Crystals*, Edited by Demus, D., Goodby, J., Gray, G.W., Spiess, H.-W., Vill, V. (Wiley-VCH, Weinhem, 1998, Vol. 1, p. 454).

Siegrist, Th., Kloc, Ch., Schoen, J.H., Batlogg, B., Haddon, R.C., Berg, S., Thomas, G.A. (2001) *Angew. Chem. Int. Ed.* **40**, 1732.

Simonescu, C., Oprea, C. *Mechanochemistry of Macromolecular Compounds* (NTIS, Springfield, VA, 1970).

Suzuki, E., Kobayashi, Yu., Endo, Sh., Deguchi, K., Kikegawa, T. (2003) *Ferroelectrics* **285**, 167.

Turner, J. (2002) *Mater. World* **10**, 26.

Urata, Sh., Irisawa, J., Takada, A., Shinoda, W., Tsuzuki, S., Mikami, M. (2005) *J. Phys. Chem. B* **109**, 4269.

van Beek, J.D., Hess, S., Vollrath, F., Meier, B.H. (2002) *Proc. Natl. Acad. Sci. USA* **99**, 10266.

Vekteris, V., Murchaver, A. (1995) *Lubr. Eng.* **51**, 851.

Wang, C., Tam, K.C., Tan, C.B. (2005) *Langmuir* **21**, 4191.

Wasem, J.V., LaMarche, B.L., Langford, S.C., Dickinson, J.T. (2003) *J. Appl. Phys.* **93**, 2202.

Wasilewski, T., Sulek, W. (2003) *Tribologia* **34**, 115.

Wazynska, B., Okowiak, Ju., Sulek, M.W. (2004) *Tribologia* **35**, 301.

Wei, K., Ho, J. (1997) *Macromolecules* **30**, 1587.

Wolter, A.U.B., Klauss, H.-H., Litterst, F.J., Burghardt, T., Eichler, A., Feyerherm, R., Sullow, S. (2003) *Polyhedron* **22**, 2139.

Yan, X.P., Perry, S.S., Spencer, N.D., Pasche, S., De Paul, S.M., Textor, M., Lim, M.S. (2004) *Langmuir* **20**, 423.

Zheng, Y., Lin, Z., Zakin, J.L., Talmon, Y., Davis, H.T., Scriven, L.E. (2000) *J. Phys. Chem. B* **104**, 5363.

6 Nano- and Biolubrication

6.1 INTRODUCTION

This chapter considers organic compounds and composites in their applications to nano- and biolubrication. Friction of mutually sliding surfaces is an understandable example from mechanodynamics. Regarding biomechanics, with each step taken by a person, slight friction takes place in a hip joint. The body relies on many other moving parts: the beating heart, chewing teeth, and blinking eyes. All the parts obey laws of biotribology. One of the specific aspects of biotribology is *biocompatibility*, which refers to the compatibility of biomaterials with the biological systems. Fully compatible bioengineering materials are unattainable, similar to the unattainability of medicines with absolutely no side effects. In biotribology, physiological tolerance is a target.

Of course, nano- and biolubrication are considered from the organic mechanochemistry point of view. Features of acting forces peculiar to nanosystems and living organisms are outside the scope of this book. The molecular transformations in the nano- and biosystems, however, can be understood from the common sense aspects of chemistry.

6.2 ANTIFRICTION AND ANTIWEAR NANOLAYERS

Particles in the nanoscale (*nanoparticles*) are isolated species with diameters well below 100 nm. Usually, hard coatings are employed of contacts to prevent mechanical deformation in the area between mating surfaces — in other words to combat wear.

An example of nanolayer lubricity stems from description of fullerenes as ball bearings (see Chapter 3). These molecular wheels work similar to wheels in the macroscopic world, in which they replace sliding high-friction motion by smooth rolling. Fullerenes as antifriction agents bridge the gap between macrotribology and nanotribology. In premature works, C_{60} films were used as solid-lubricating coating (Bhushan et al. 1993a, 1993b). Then, fullerene C_{60} was studied as an additive to liquid oils (Gupta and Bhushan 1994). The results were encouraging. According to Ginzburg, Shibaev, and coauthors (2005), hydrocarbon components of the oil generate free radicals during friction. Fullerene accepts these radicals, acting as a center of the three-dimensional polymeric network. (Each molecule of C_{60} can add up to six macroradical chains; Hirsh 1998.) This type of triboreaction led to the formation of a fullerene with six arms grafted to it. Such nanoparticles eventually transform in the wear-resisting films. These films markedly reduce friction and enhance antiwear properties and high-pressure workability of the lubricating oils. For the oils, the running-in period is significantly shortened. In the presence of fullerene (as the pure compound or as part of soot), the work life of the specific lubricating oil is increased even if antifriction and antiwear properties of the oil are not improved (Ginzburg et al. 2004; Ginzburg, Kireenko, et al. 2005).

Radical-initiated copolymerization of fullerene C_{60} with acrylamide leads to the formation of nanoballs with an average diameter of 46 nm. The core of the copolymer is a very hard fullerene; the shell is polyacrylamide, which is relatively soft but very elastic. The addition of 0.2% of the copolymer to base stock (2% triethanolamine aqueous solution) effectively raises the load-carrying capacity and antiwear ability (Jiang et al. 2005). Triethanolamine is needed to decrease the corrosive activity of pure water. The solution modified with a small amount of fullerene-contained copolymer can be used as an improved metal-working fluid.

Fulleroids are even more effective in nanolubrication. They are obtained alongside fullerenes when the graphite anode is thermally pulverized at lowered pressure in the inert atmosphere. Fulleroids are polyhedral multilayer nanostructures that differ from fullerenes by higher thermal stability and hardness. Fulleroid nanomodifiers exert a positive influence on the structure, the strength, and the working durability of matching parts. Optimally, a hundredth or even a thousandth of 1% is sufficient to markedly decrease wear of mechanical parts (Blank et al. 2003).

A novel lubrication strategy was proposed for micromechanical devices. The strategy implies the formation of elastomeric boundary lubricating coatings from block copolymers grafted to a reactive surface (Tsukruk 2001). Silicon oxide substrate was used as the reactive surface; the elastomeric layer was poly[styrene-*b*-(ethylene-*co*-butylene)-*b*-styrene] functionalized with maleic anhydride. This material formed the nanophase-separated domain structure in the bulk. The epoxy-terminated self-assembled monolayer fabricated from epoxyalkylsilane was used as a reactive anchoring interface on a silicon oxide wafer. The maleic anhydride fragments of the elastic block reacted with the epoxy group of the monolayer, thus enabling anchoring of the elastic block to the surface. In assembly, the system contains the following three layers: (1) an elastomeric (rubber) layer with very low shear modulus; (2) a reinforced nanodomain layer with much higher compression modulus, and (3) a reactive interfacial layer between the rubber layer and the modified surface. The reactive interfacial layer serves as an anchor and a buffer between the main elastomeric layer and the solid surface. The solid surface was the silicon oxide layer (as a part of the micromechanical system). Such triple-layer film is extremely wear resistant under shear stress because of its unique combination of a chemically grafted elastic matrix capable of tremendous reversible deformation (500–1000%) and stabilized by a chemically interconnected nanodomain net. In the case of complete polymer films with well-developed nanodomain microstructures, the friction coefficient dropped to a low value, below 0.03 (Tsukruk 2001).

6.3 BIOTRIBOLOGY

6.3.1 Lubrication in Natural Joints

The surface of each bone is not in direct contact with another, but is covered by a special articular cartilage that resembles the hyaline cartilages found elsewhere in the body. Synovia is lubricating fluid secreted by the membranes of animal and human joint cavities, tendon sheaths, and the like. *Synovia* comes from the Greek *syn* (with) and the Latin *ovum* (egg). Synovial fluid ensures the abnormally low friction of natural joints in human and animal organisms. As proven (Kupchinov et al. 2002), this fluid contains endogenic cholesterol compounds that are retained within the joint bursa, a sac or pouchlike cavity between joints or between tendon and bone.

Synovial fluid originates from the blood plasma and has a composition similar to plasma. The plasma is composed of cholesterol esters. In the zone of friction of the joint cartilage, these esters produce a liquid crystalline nematic phase. Within the interval of physiological temperatures, the liquid crystal phase exists permanently. Synovial liquids taken from kneecaps of human donors (28 men and 18 women) or neat cattle (15 animals) were analyzed by gas chromatography. The following cholesterol esters were identified: hexadecanoate (palmitate), 7-hexadecynoate (palmitoleate), octadecanoate (stearate), butanoate (oleate), 9,12,15-octadecatrienoate (linoleate), and 5,8,11,14-eicosatetraenoate (arachidonate) (Kupchinov et al. 2002). As seen, all the esters belong to the class of long-chain acid derivatives. Because they are attached to the cholesterol moiety, they form cigarlike molecules, which is a typical shape of liquid crystals.

Addition of the esters to a test lubrication solution (sodium salt of carboxymethyl cellulose in water) provides characteristics of cartilage friction equal to those of the natural synovial liquid. The higher the ester content is, the lower is the friction (Kupchinov et al. 2002).

To date, hyaluronic acid (mucopolysaccharide) was considered the main lubricating component of biojoints (Mow et al. 1990). This component is also present in synovial liquid. At the same time, it is known that selective enzymatic depolymerization of this mucopolysaccharide does not practically affect the lubricity of hinges (Wright et al. 1973). Supposedly, the cholesteric liquid crystalline layers orient the hyaluronic acid molecules along the crystal long axes according to the guest–host mechanism. However, the antifriction action springs from the cholesterol esters. Such inference opens a new line in search of medicines capable of modifying properties of endogenic cholesterol compounds when some pathology takes place in the natural joints.

Another line in such searches suggests use of dipalmotoyl phosphatidyl choline in propylene glycol solution in injections to reduce both friction and wear in osteoarthritic natural joints. This choline derivative is a constituent of synovial fluid and differs in its high surface activity and capability to deposit onto natural joint asperities (Jones et al. 2004; Ozturk et al. 2004).

6.3.2 Lubrication in Artificial Joints

As the average life expectancy in much of the world increases, there is a concomitant increase in the need for replacement body parts. More than 200,000 people per year in the United States receive a hip prosthesis. Despite modern materials used for the implants, statistical data show that after 15 years, 15–20% of these devices will have failed (Widmer et al. 2001). This limitation in the lifetime is mainly because of wear of prosthetic materials, and it represents a serious and urgent problem for organic mechanochemistry. Artificial joints require lubricating compositions that differ from natural articular lubricants. Even if the latter go in the prostheses they cannot ensure the work of artificial joints. Because of the solid-to-solid contact, friction and wear generate wear particles.

The wear itself causes loosening of the prosthesis. If the wear becomes extensive, the artificial hip needs to be replaced. The wear particle, however, is of much greater concern. Wear debris can migrate to distant organs, particularly the lymph nodes, where accumulation of particle-containing macrophages cause chronic lymphadenitis. The body's immune system attempts, unsuccessfully, to digest the wear particles (as if it were a bacterium or virus). Enzymes are released. This eventually results in the death of adjacent bone cells (*osteolysis*). Over time, sufficient bone is resorbed around the implant to cause mechanical loosening, which necessitates a costly and painful implant replacement.

In 1983, the poly(tetrafluoroethylene) (PTFE) implant was approved for the market. As pointed out in Chapter 5, this polymer exhibits a low coefficient of friction. However, of the more than 25,000 PTFE implants received by patients, most failed. Because PTFE has unacceptable wear properties, it was replaced with ultrahigh molecular weight poly(ethylene) (UHMWPE), which has a much lower wear rate. This polymer has long been the material of choice for bearing surfaces used in total hip and knee athroplasty.

Considering the small volume of lubricant contained in an articular capsule and the poor thermal properties of polymers, frictional heating is definitely related to the wear of UHMWPE through softening. Bergmann et al. (2001a, 2001b) reported that the *in vivo* temperature of femoral heads rose more than 43°C after an hour of walking. Therefore, the potential risk of thermal damage on the stability of hip implants cannot be excluded. According to these authors, the nature of the implant material plays a decisive role in dissipation of the frictional temperature effect. There are also other effects specific for orthopedic applications of UHMWPE. Modern design of hip prostheses is usually based on use of cobalt or titanium vanadium aluminum alloy and UHMWPE Chirulen, which is a trade name for UHUWPE manufacturing by Gsell Engineering plastics AG (Switzerland). The friction pair of such an endoprosthesis consists of the alloy head and polyethylene cup. One of the shortcomings is insufficient tribological properties of the device.

Increasing the resistance of UHMWPE to wear and damage was proposed by means of radiation cross-linking and subsequent melting (Muratoglu and Scholar 2004; Oonishi et al. 2003) (see the next paragraph). Cross-linking improves the wear resistance of this polymer; postirradiation melting improves the long-term oxidative stability, which is the primary precursor to polyethylene damage *in vivo*. Molding of polyethylene with Kevlar (polyamide fibers) seemingly improves its mechanical properties in acetabular cups tested *in vitro* and makes it much more biocompatible (Chowdhury et al. 2004). Kevlar [poly(paraphenylene terephthalamide)] is a Dupont product (Delaware, U.S.A)

Promising results were obtained with fullerene C_{60}-based materials as coatings of the wearing surface of endoprostheses (Lashneva et al. 2003). The titanium alloy with fullerene C_{60} coatings has approximately the same wear resistance as polished alumina ceramics and about tenfold more wear resistance than the same alloy without coating. The wear at [Chirulen]-[Ti(6)Al(4)V alloy coated by C_{60}] friction pair was ten times less than that at the same Chirulen–alloy pair but without the fullerene coating and was comparable to Al_2O_3 medical ceramics paired with Chirulen. After friction for 20 h, the surface roughness of [Ti(6)Al(4)V] disks coated with Chirulen and C_{60} were almost the same as before the friction. There were no surface scratches, alteration of color, or diminution of the thickness of the alloy. If clinical tests are successful, the titanium alloys with fullerene coatings can be used in endoprosthesis friction pairs of either upper or lower limb joints. (Whereas the upper limbs do not need to carry body weight during motion, the lower limbs are under the load of the whole body. However, the upper limbs sometimes receive high mechanical stress and must be load resistant.)

6.3.3 REDOX REACTION PROBLEMS OF ARTICULATE BIOENGINEERING

Whereas cross-linking is an understandable method for UHMWPE reinforcement, the role of its postirradiation melting should be discussed in more detail. During irradiation, free radicals are formed. Within the crystalline domains, they are not able to recombine. These residual free radicals further lead to the oxidation and embrittlement of polyethylene. They can readily react with oxygen and form peroxy free radicals. The peroxy radicals transform into hydroperoxides by abstracting hydrogen atoms from nearby carbon atoms. The abstraction of hydrogen produces a new free-radical center, which in turn takes part in the oxidation cascade. The hydroperoxides are unstable and decay into carbonyl species, mainly ketones and acids. This reduces the molecular weight of the polymer, leading to recrystallization and an increase in stiffness and embrittlement. Therefore, it is necessary to eliminate the residual free radicals formed as a result of radiation cross-linking. The most effective method is to raise the temperature of polyethylene above its peak melting transition (about 137°C) to eliminate the crystalline domains and liberate the trapped residual free radicals.

Radiation cross-linking and melting increase the wear and oxidation resistance of UHMWPE. At the same time, however, the fatigue resistance is diminished. A fatigue crack starts and propagates when the localized stress at the crack tip cannot be dissipated by energy absorption within the regions ahead of the tip. Plastic deformation of the crystalline domains plays a decisive role in energy dissipation. Such a type of plastic deformation (so-called crystal plasticity) depends on the ductility and crystallinity of the material. Understandably, cross-linking reduces the chain mobility of UHMWPE, decreasing its ductility. Postirradiation melting decreases the crystallinity. In total, resistance of the material against fatigue declines. To avoid the necessity of postirradiation melting, Oral and colleagues (2004) proposed treating a highly cross-linked UHMWPE with the commercially available antioxidant D-α-tocopherol (also known as vitamin E). Oxidation and wear resistance of the material, which was not molten but enriched with tocopherol, was comparable to contemporary highly cross-linked/melted UHMWPE and exceeded the latter in fatigue resistance. The results opened new possibilities for creation of high-stress joint implants with considerably increased longevity.

UHMWPE is paired with metallic parts of artificial joints. Stainless steel paired with polyethylene produces higher wear rates than the cobalt–chromium alloy with polyethylene. In turn, the cobalt–chromium alloys showed superior wear resistance

12 years after the implantation surgery. The titanium alloys were significantly damaged before 10 years *in vivo* use (Brummitt et al. 1996). For hip replacement, femoral
heads manufactured of aluminum oxide or zirconium oxide ceramics were fine when
articulating with polyethylene. The ceramic–polyethylene possessed the highest wear
resistance compared to metal alloys (Urban et al. 2001). In common opinion, however, the ceramic materials are prone to fracture. In practice, the failure rate of
ceramic femoral heads reached 13.4% in 1980–1990. Such a rate forced altered
technology of head manufacture and thereby lowered the failure rate to 0–2%
(Davidson 1993). The ceramic–polyethylene pair is the most attractive material for
joint implants at the present time. Beneficial properties of ceramics as an implant
material include their chemical inertness, lack of solubility, smooth surface, and
hydrophilic character. The smooth wetted ceramic surface produces lower coefficients of friction compared to conventional implant metals.

One additional (and unique) merit of ceramic materials is that, in contrast to
metal alloys, ceramics do not form an oxidative coating that leads to oxidative wear.
The biological environment is rich with oxygen. Unlike ceramics, implant metal
reacts with oxygen to form oxides that provide the metal surface with a protective
coating that prevents corrosion. The oxide film forms instantly once exposed to *in
vivo* conditions but can be scratched or rubbed off during motion. Formation and
removal of the oxide film repeat cyclically. As a result, the implant metal releases
metal ions and very small, separate particles. In the situation termed *third body wear*,
roughness is substantially increased. Accordingly, wear rates increase dramatically.
The whole process is termed *oxidative wear*. Although metallic implants show such
release, ceramic parts are inert in this sense.

Nevertheless, transition metal alloys are still used orthopedically. Transition
metals usually generate cations in the lower oxidation state, say Co(II), not Co(III).
In aqueous solutions, transformation of Co(II) to Co(III) is characterized by a high
enough redox potential of +1.8 V. This precludes one-electron transfer from Co(II)
to O_2 with formation of Co(III) and O_2^{-}. However, expelling Co(II) in the biological
environment containing biomolecules puts some chemical subtleties forward. Complexation of Co(II) with amino acids or their bearers proceeds readily and changes
the metal redox potential. Such potential of cobalt triglycinate $[Co^{II}(Gly)_3]^-$ is +0.20 V
(Hin-Fat and Higginson 1967). This makes superoxide ion generation possible
according to the reaction $[Co^{II}(Gly)_3]^- + O_2 \rightarrow [Co^{II}(Gly)_3 + O_2^{-}$ (Silwood et al.
2004). Superoxide ion induces many malignant effects in living organisms and
even provokes cancer.

Similarly, redox-active Ti(III) species is released from titanium metal prostheses.
In the knee joint sinovial fluid, Ti(III) transforms into Ti(IV) with simultaneous
generation of superoxide ion. High-field ^1H nuclear magnetic resonance spectroscopy indicated that Ti(IV)–citrate complexes are present in this biofluid (Silwood
and Grootveld 2005).

Of course, there are other reactions with endogenic bioorganics that lead to
oxidative damage of organic implants. The corrosive liquids of the knee or hip
environment produce chemical oxidative processes, so-called biological wear. Using
diverse physical methods, Torrisi et al. (2004) studied the unicompartmental knee
prostheses of woman patients more than 70 years old who were explanted after

3 years of insertion. These prostheses were representative of the general wear behavior *in vivo*. Biological action generally acts by increasing mechanical degradation. In other words, biological wear, present already after a few years, catalyzes the mechanical wear. Mechanical and biological wear changes the physical properties of the material, such as roughness and hardness in the stressed area. The study of these two parameters on all the polymeric matches made it possible for the authors to identify the area stressed, which was about 1 cm^2. When implanted with heavy ions (e.g., Xe$^+$), UHMWPE sheets prepared in the laboratory had enhanced mechanical resistance. Torrisi and coauthors (2004) suggested that such heavy ion implantation can be successfully applied to polymer implants. The main effect is polymer surface modification in terms of increased hardness and improved wear resistance.

6.3.4 Innovations in Organic Materials for Articulate Prostheses

In continuing our consideration of organic mechanochemical reactions, we should not enforce the sometimes-noted pessimism regarding arthroplactics as a whole. Practitioners choose appropriate materials and pharmaceutical supporting methods that assist bone on-growth and in-growth. The target is that the utility of artificial joints will outlive the host's life span.

As an example of pharmaceutical supporting material, a new generation of phosphorus-containing compounds can be adduced. Novartis Pharma AG (Basel, Switzerland) offers [1-hydroxy-2-(1H-imidazol-1-yl)ethylidene] biphosphonic acid to decrease polyethylene particle-induced osteolysis. To be effective, preceding generations of biphosphonates had to be administered daily. The imidazole-containing biphosphonic acid is more potent; it can be used as a single dose introduced beneath the skin (von Knoch et al. 2005). This important new method of drug administration holds great promise because single-dose treatment of particle-induced osteolysis may reduce side effects compared to repeated application of biphosphonate.

Seeking new materials, Howling and coworkers (2003) observed that P25 carbon fibers (Goodfellow, Cambridge, England) conjugated with chemically deposited methane had a very low wear factor and generated particles that were extremely small. Principally, the same behavior was fixed for joints made from short polyacrylonitrile-based fibers with a primary matrix of resin-coke and a secondary matrix of pitch (Howling and coworkers 2004). Stoy (2004) disclosed a method of expedient formation of lubricious polyacrylonitrile artificial joints available for bone regeneration. What is especially promising that is a very low level of debris produced from the joints during their work. Furthermore, the debris size is less than 100 nm, which is much smaller than that of polyethylene wear particles. Such small wear particles do not activate inflammatory cells and therefore produce fewer osteolytic reactions.

Poly(vinyl alcohol)-hydrogel is also a promising material for hemiarthroplastic surgery (Kobayashi et al. 2005; Kobayashi and Oka 2004). Hydrogels are hydrates of hydrophilic macromolecular chains. Hemiarthroplastic surgery consists of the treatment of hip joint disorders in which the lesion is limited to the joint surface. The hydrogel behaves like a highly viscous fluid. This is a highly dense colloid gel. Its lubricating activity is similar to that in the joint (Ishikawa and Sasada 2004).

6.3.5 ORAL LUBRICATION

Salivary proteins play an essential role in oral lubrication, which in turn is important for maintaining such functions as tissue protection and speaking, mastication, and deglutition capabilities. Many people suffer from impaired salivary function, displaying various symptoms, such as abnormal wear of the dentition and "dry mouth" (xerostomia). Thus, there is a need for saliva substitutes that mimic the lubricating properties of native saliva. Commercially available saliva substitutes were enumerated in publications by Hatton et al. (1987) and Reeh et al. (1996). It was concluded that the presence of salivarylike pellicles on hard surfaces alters force behavior. The friction coefficient is largely reduced in accordance with the purely repulsive long-range force acting between the films.

The majority of commercial medicines administered to patients with dry mouth cause marked side effects. To treat reduction or loss of salivary production, a bioadhesive gel was formulated for localized treatment (Kelly et al. 2004). An important distinguishing characteristic of the gel is its prolonged residence time. The formulation is based on the polymer Carbopol 974P (Reprotect LLC, Baltimore, MD, U.S.A.). To enhance lubricity and taste of the product, sunflower oil was introduced. The final formulation also contains salivary levels of electrolytes to help remineralization of teeth, fluoride to prevent caries, zinc to enhance taste sensation, triclosan as the main antimicrobial/anti-inflammatory agent, and noncariogenic sweeteners with lemon flavor to increase the palatability of the product. The gel stimulates residual salivary function.

6.3.6 OCULAR TRIBOLOGY

Tears are required for normal functioning of visual systems. Actually, tears are composed of lipid, aqueous, and mucin components. The tear film provides a smooth surface for light refraction and plays an important role in the ocular defense system. What is especially relevant to our consideration is the tear lubricating function for eyelids, the conjunctiva, and the cornea. Eyelids are the movable folds of flesh that cover and uncover the front of the eyeball. Conjunctiva are the mucous membranes lining the inner surface of the eyelids and covering the front part of the eyeball. The cornea is the transparent tissue that forms the outer coat of the eyeball and covers the iris and pupil. All of the parts require effective lubrication to prevent damage to the sensitive ocular system. Deficiency in the amount of tear production or alteration in tear composition can lead to ocular pathology. Dry eye syndromes are frequently encountered. To treat these syndromes, artificial tears must be formulated and introduced. The ideal tear replacement should have a composition that is compatible with the maintenance of a normal ocular surface epithelium. Furthermore, it should be able, when damage of the ocular surface exists, to provide an environment in which the epithelium could recover normal structure and function. Despite extensive research in the field, the major problem in the ocular drug delivery domain still is rapid precorneal drug loss and poor corneal permeability. Because of a decline in the quality or quantity of the tear fluid, these syndromes affect 10 million people in the United States.

Tears are a complex combination of substances that form three layers on the eye. The highly viscous outer 0.1-μm layer is composed mainly of fatty molecules (lipids). The aqueous layer is in the middle. It has low viscosity and is 10 μm thick.

The innermost layer is composed of viscous mucus that helps the tear film stick to the surface of the eye.

With each blink, the tear film is re-formed, so it needs to be stable when eyes are open between blinks. Tear film stability depends on its physical properties, particularly surface tension and viscosity. The dynamic surface activity of most biological fluids, including tears, is a result of a spatially specific interaction between lipids and proteins. In a result, the lipid is oriented by the protein into a preassembled lamellar form that can be readily absorbed at the biological interface. Lack of such assemblies also causes the lubricity deficiency diseases, of which dry eye syndrome in one example. Such assemblage, in other words such nanostructure, is realized in the mixture of dilauroyl phosphatidyl choline with the styrene–maleic acid alternating copolymer (Tonge et al. 2002; Tonge and Tighe 2002). The authors proposed this therapeutic agent to circumvent dry eye syndrome.

A method was claimed to alleviate the symptoms of dry eye by administrating to the eye a diluted aqueous solution of hydroxypropyl methylcellulose and hydroxyethyl cellulose (Chowhan and Chen 2004). The solution does not form a gel. The presence of two polymers significantly enhances the viscosity and lubrication property of a composition while minimizing total polymer concentration and cost of materials. The composition is suitable for use as artificial tears or as a vehicle for ophthalmic drugs.

Another new lubricating eye drop contains two demulcents (polyethylene glycol 400 and propylene glycol) with hydroxypropyl guar as a gelling agent. This medication is marketed in the United States according to the Ophthalmic Drug Products Permission for human use. Clinical tests demonstrated both the safety and the efficacy of the medication. The benefit of the treatment is even greater among patients with severe pretreated ocular problems (Christensen et al. 2004). From a chemical point of view, the hydroxypropyl guar component deserves special consideration. It is a derivative of guar gum, a high molecular weight branched polymer of mannose and galactose in a 2:1 ratio from the guar bean (*Cyanopsis tetragonoloba*). This polymeric system is a liquid in the bottle at pH 7.0, but forms a soft gel when exposed to the approximate pH 7.5 of the tear film. When the formulation is exposed to the patient's tears, it generates an increase in viscosity, producing a mucinlike coating on the ocular surface. This provides a long-term protective layer that prevents desiccation and promotes recovery of the damaged epithelium (Ubels et al. 2004).

6.4   CONCLUSION

For artificial hip and knee implants, the materials comprising the sliding surfaces are very different from the cartilage surfaces that slide in the natural joints. The commonly exploited UHMWPE sliding surface is hydrophobic. It interacts with proteins existing in the joint space, the space that is a part of the physiological medium. Hydrophobic surfaces, although generally absorbing proteins strongly, also have a tendency to denaturize them. This decreases the ability of protein to participate in the lubrication phenomena. Of course, hydrophilic surfaces are discernible in their hydrophilicity. When the surfaces are more hydrophilic, protein denaturation decreases. In this case, proteins remain in a more effective state for boundary lubrication (Widmer et al. 2001). In other words, there is a need for new biocompatible polymers with greater surface hydrophilicity (say, at the expense of incorporating polar functionalities in the polyethylene backbone). Regarding inorganic counterparts of the artificial joint, the ceramic countersurface is also hydrophilic. Indeed, proteins are adsorbed in their natural form on ceramics (Heuberger et al. 2005). Zirconia lubrication also benefits from proteins, although the mechanism of this effect remains unknown (Clarke et al. 2003).

Introducing antioxidants into artificial articular cartilage seems to be one effective way to enhance the antifragile properties of the material. Cross-linked UHMWPE can be very attractive in this sense, but only when an antioxidant is introduced into the whole mass of the polymer and if its inside content is on a proper level during the work of the orthopedic joint.

Regarding ophthalmology problems, Dr. Frank J. Holly (the president of the Dry Eye Institute in Lubbock, TX) dared say that the current practice of using high-viscosity lubricants or ointments to treat eye problems may actually interfere with lid lubrication and tear formation (see Jacobson 2003). Holly argued that, when the eye closes during a blink, the outer lipid layer is compressed and swept away by the moving eyelid. So, the lubricating layer between the lid and the eye is actually the aqueous tear layer, bounded on both sides by mucus. The mechanism of eye lubrication therefore is hydrodynamic. In Dr. Holly's opinion, less-viscous lubricants would work better. This means that we can wait for new, more effective, formulations to circumvent dry eye syndromes.

When therapeutic courses allow postponing surgical invasion, knowledge of mechanochemical principles helps create an appropriate treatment for patients. Chapter 6 gives a number of examples for such an approach.

REFERENCES

Bergmann, G., Graichen, F., Rohlmann, A., Verdonschot, N., van Lenthe, G.H. (2001a) *J. Biomech.* **34**, 421.

Bergmann, G., Graichen, F., Rohlmann, A., Verdonschot, N., van Lenthe, G.H. (2001b) *J. Biomech.* **34**, 429.

Bhushan, B., Gupta, B.K., Van Cleef, G.W., Capp, C., Coe, J.V. (1993a) *Appl. Phys. Lett.* **62**, 3253.

Bhushan, B., Gupta, B.K., Van Cleef, G.W., Capp, C., Coe, J.V. (1993b) *Tribol. Trans.* **36**, 573.

Blank, E.D., Vinogradov, S.E., Oryshchenko, A.S., Rybin, V.V., Slepnev, V.N., Shekalov, V.I., Chernigovskii, A.A. (2003) *Vopr. Materialovedeniya* **3**, 65.

Brummitt, K., Hardaker, C.S., McCullgh, P.J., Drabu, K.J., Smith, R.A. (1996) *J. Eng. Med.* **210**, 191.

Chowdhury, S.K.R., Mishra, A., Pradhan, B., Saha, D. (2004) *Wear* **256**, 1026.

Chowhan, M., Chen, H. (2004) *U.S. Pat.* 0253280.

Christensen, M.T., Cohen, S., Rinehart, J., Akers, F., Pemberton, B., Bloomenstein, M., Lesher, M., Kaplan, D., Meadows, D., Meuse, P., Hearn, Ch., Stein, J.M. (2004) *Curr. Eye Res.* **28**, 55.

Clarke, I., Green, D.D., Pezzotti, G., Sakakura, S., Ben-Nissan, B. (2003) *Ceram. Trans.* **147**, *Bioceram.: Mater. Appl. IV*, 155.

Davidson, J. (1993) *Clin. Orthoped. Rel. Res.* **294**, 361.

Ginzburg, B.M., Kireenko, O.F., Shepelevskii, A.A., Shibaev, L.A., Tochil'nikov, D.G., Leksovskii, A.M. (2005) *J. Macromol. Sci., Phys.* **44**, 93.

Ginzburg, B.M., Shibaev, L.A., Kireenko, O.F., Shepelevskii, A.A., Melenevskaya, E.Yu., Ugolkov, V.L. (2005) *Vysokomol. Soedin., Ser. A, Ser. B* **47**, 296.

Ginzburg, B.M., Shibaev, L.A., Melenevskaya, E.Yu., Pozdnyakov, A.O., Pozdnyakov, O.E., Ugolkov, V.L., Sidorovich, A.V., Smirnov, A.S., Leksovskii, A.M. (2004) *J. Macromol. Sci., Phys.* **43**, 1193.

Gupta, B.K., Bhushan, B. (1994) *Lubr. Eng.* **50**, 524.

Hatton, M.N., Levine, M.J., Margarone, J.E., Aguirre, A. (1987) *J. Oral Maxilofac. Surg.* **45**, 496.

Heuberger, M.P., Widmer, M.R., Zobeley, E., Glockshuber, R., Spencer, N.D. (2005) *Biomaterials* **26**, 1165.

Hin-Fat, J., Higginson, W.C.E. (1967) *J. Chem. Soc. A*, 298.

Hirsch, A. In: *Fullerenes and Related Structures. Topics in Current Chemistry*. Edited by Hirsch, A. (Springer-Verlag, Berlin, Germany, 1998, p.1).

Howling, G.I., Ingham, E., Sakoda, H., Stewart, T.D., Fisher, J., Antonarulrajah, A., Appleyard, S., Rand, B. (2004) *J. Mater. Sci.: Mater. Med.* **15**, 91.

Howling, G.I., Sakoda, H., Antonarulrajah, A., Marrs, H., Stewart, T.D., Appleyard, S., Rand, B., Fisher, J., Ingham, E. (2003) *J. Biomed. Mater. Res., Part B: Appl. Biomater.* **67B**, 758.

Ishikawa, Y., Sasada, T. (2004) *Mater. Trans.* **45**, 1041.

Jacobson, A. (2003) *Tribol. Lubr. Technol.* **59**, 34.

Jiang, G.-Ch., Guan, W.-Ch., Zheng, Q.-X. (2005) *Wear* **258**, 1625.

Jones, C.F., Stoffel, K.K., Ozturk, H.E., Stachowiak, G.M. (2004) *Tribol. Lett.* **16**, 291.

Kelly, H.M., Deasy, P.B., Busquet, M., Torrance, A.A. (2004) *Int. J. Pharm.* **278**, 391.

Kobayashi, M., Chang, Y.-Sh., Oka, M. (2005) *Biomaterials* **26**, 3243.

Kobayashi, M., Oka, M. (2004) *J. Biomater. Sci., Polym. Ed.* **15**, 741.

Kupchinov, B.I., Ermakov, S.F., Rodenkov, V.G., Beloenko. E.D., Eysmont, E.S. (2002) *Trenie Iznos* **23**, 310.

Lashneva, V.V., Tkachenko, Yu.G., Dubok, V.A., Schur, D.V., Sychev, V.V., Matveeva, L.A. (2003) *NATO Sci. Ser., II: Math., Phys., Chem.* **102** (*Nanostruct. Mater. Coating Biomed. Sensor Appl.*), 103.

Mow, V.C., Ratcliffe, A., Woo, S.L.-Y., Eds. *Biomechanics of Diarthrodial Joints* (Springer-Verlag, Berlin, Germany, 1990, vol. 2).

Muratoglu, O.K, Scholar, A.G. (2004) *62nd Annual Technical Conference — Society of Plastics Engineers,* **3**, 3744.

Oonishi, H., Clarke, I.C., Yamamoto, K., Masaoka, T., Fujisawa, A., Masuda, Sh. (2003) *J. Biomed. Mater. Res., Part A* **68**, 52.

Oral, E., Wannomae, K.K., Hawkins, N., Harris, W.H., Muratoglu, O.K. (2004) *Biomaterials* **25**, 5515.

Ozturk, H.E., Stoffel, K.K., Jones, C.F., Stachowiak, G.W. (2004) *Tribol. Lett.* **16**, 283.

Reeh, E.S., Douglas, W.H., Levine, M.J. (1996) *J. Prosthetic Dent.* **75**, 649.

Silwood, C.J.L., Chikanza, I.C., Tanner, K.E., Shelton, J.C., Bowsher, J.G., Grootveld, M. (2004) *Free Radical Res.* **38**, 561.

Silwood, C.J.L., Grootveld, M. (2005) *Biochem. Biophys. Res. Commun.* **330**, 784.

Stoy, G.P. (2004) *U.S. Pat.* 0070107.

Tonge, S., Rebeix, V., Young, R., Tighe, B. (2002) *Adv. Exper. Med. Biol.* **506** (*Lacrimal Gland, Tear Film, Dry Eye Syndromes 3, Part A*), 593.

Tonge, S.R., Tighe, B.J. (2002) *Polym. Mater. Sci. Eng.* **87**, 495.

Torrisi, L., Visco, A.M., Campo, N., Rizzo, D., Bombara, A. (2004) *Bio-Med. Mater. Eng.* **14**, 251.

Tsukruk, V.V. (2001) *Tribol. Lett.* **10**, 127.

Ubels, J.L., Clousing, D.P., Van Haitsma, T.A., Hong, B.-Sh., Stauffer, P., Asgharian, B., Meadows, D. (2004) *Curr. Eye Res.* **28**, 437.

Urban, J.A., Garvin, K.L., Boese, C.K., Bryson, L., Pedersen, D.R., Callaghan, J.J., Miller, R.K. (2001) *J. Bone Joint Surg.* **83A**, 1688.

von Knoch, M., Wedemeyer, Ch., Pingsmann, A., von Knoch, F., Hilken, G., Sprecher, Ch., Henschke, F., Barden, B., Loeer, F. (2005) *Biomaterials* **26**, 1803.

Widmer, M.R., Heuberger, M., Voros, J., Spencer, N.D. (2001) *Tribol. Lett.* **10**, 111.

Wright, V., Dowson, D., Kerr, J. (1973) *Int. Rev. Connect. Tissue Res.* **6**, 105.

7 Concluding Remarks and Horizons

7.1 INTRODUCTION

The synthetic advantages of mechanically initiated organic reactions are widely demonstrated in this book. We are on the eve of a new era in organic chemistry — the era when the dominating role of reaction solvents recedes. This refers to both laboratory synthetic methods and organic chemistry manufacturing. The pharmaceutical industry provides excellent examples of such an approach. Public chemophobia gradually will become groundless.

Because the very tenor of the book concerns practical applications of mechanically induced organic reactions, their industrial significance is underlined and disclosed. The same approach is employed for those reactions and substances that can be predicted as important in the near future. Not all aspects of organic reactions on mechanical stress are completely understood. The majority of the references provided are recent. Observe that new interpretation of scientific data appears frequently. I have attempted to synthesize the ideas from various references that complement one another, although the connections among them may not be immediately obvious. For this reason, an author index is included to help find such connections in the book. I apologize to any authors who have contributed to the development of this vast field but, for various reasons, have not been cited. The contributors who are cited certainly do not reflect my preferences; their publications have been selected as illustrative examples that may allow the reader to follow the evolution of the corresponding topics.

Some important points from this book are concisely expressed here to underline their scientific and practical significance.

7.2 MECHANOCHROMISM AND INFORMATION RECORDING

Important experimental work some years ago by Zink et al. (1976) led the authors to suggest a rule concerning the occurrence of triboluminescent activity in crystalline compounds. Namely, only the structures that lacked inversion symmetry, commonly called noncentric crystals, could display triboluminescence. The work gave a rationale for this plausible idea (see also references in Zink et al.'s 1976 article). The new works discussed in Chapter 2, however, weaken the rigor of the link between triboluminescence activities and the noncentric crystallinity of organometallic complexes. This widens a set of possible candidates for devices that record information and material damage sensors.

7.3 LUBRICITY MECHANISM AND LUBRICANT DESIGN

As many, various scientific works exist on additive tribomechanics, I mostly focused on tribochemistry. Chapter 3 concentrates on the chemical film formation and depletion associated with lubricant degradation. Chemical reactions on the rubbing surfaces control the efficacy of the lubrication process. If the chemical film fails, then the lubrication process fails. Under boundary lubrication conditions, complicated chemical processes occur at the asperity tip contacts. A chemical reaction takes place between the lubricant film and the iron (or other metal) surface. This film is likely to be composed of frictional polymers (organometallic compounds) adsorbed on the asperity tips, which provide lubrication by forming an easily shearable layer on the rubbing surfaces.

There are two processes taking place simultaneously: polymerization to increase the molecular weight and dynamic shearing from the rubbing contact to reduce the molecular weight. The formation and depletion rates of such films are associated with the lubricant degradation process. Because the chemical reactions strongly depend on temperature, the warming effects within the contact zone appear to be critical for control of the effectiveness of boundary lubrication. Many factors should be taken into consideration. They are the asperity temperatures within the contact zone of two rough surfaces, experimental conditions such as sample geometries, load, speed, material elastic constants, lubricant properties, friction coefficients, and surface roughness profiles of the worn samples.

Tribopolymerization is one newly developed and very promising branch of tribochemistry. Tribopolymerization involves the continuous formation of thin polymeric films on rubbing surfaces at the expense of monomers introduced into initial lubricating oil. The protective films formed are self-replenishing. The formation of the protective films proceeds, for instance, by means of polycondensation according to the following chemical equation: $HOOCRCOOCH_2CH_2OH \rightarrow HO[OCRCOOCH_2CH_2O]_nH$. The reaction takes place in paraffin mineral oil. For the formulation corresponding to 1% of this poly(monoester) in the oil, the wear rate is reduced by over 90% in the case of automotive engine lubrication. The compositions may also include lactam condensation monomers, which form polyamide films on the rubbing surface (Furey et al. 2002). These authors underlined that the antiwear compounds developed as a result of tribopolymerization are effective for significant reduction of friction between metals and ceramics. Furthermore, such tribofilms are ashless and contain no harmful phosphorus or sulfur, and many are biodegradable. Their would-be applications are diverse and have cost, performance, energy, and environmental advantages. Tribopolymerization opens the door to new classes of additives for "minimalist" boundary lubrication and thus offers more options and possibilities to reduce pollution.

One ripening problem in lubrication engineering is the replacement of steel by aluminum to reduce vehicle weight. This is one of the routes taken to improve fuel consumption. For each 10% reduction in vehicle weight, fuel economy could be improved by 7%. However, zinc dialkyldithiophosphate (the main lubricant of worldwide distribution) is unable to adequately protect an aluminum surface. Therefore, automobile manufacturers have resorted to engines composed of aluminum-based composite materials or to engine blocks that contain steel sleeves. These measures are costly and complicate engine fabrication. Practice requires formulations effectively protect aluminum surfaces from wear and friction in various conditions. Some solutions to this important problem are discussed within the present book.

7.4 SPECIFIC SYNTHETIC OPPORTUNITIES OF SOLVENT-FREE REACTIONS

When solvents are used, cost and environmental problems emerge. Solvents are high on the list of harmful chemicals for two reasons: they are used in large amounts, and they are usually volatile liquids dangerous in the sense of flammability and explosiveness. As a rule, mechanochemical reactions do not need any solvent. The advantages of these procedures are efficiency and their economical and environmentally benign nature. Of course, they are power consuming. However, solvent removal from the product obtained also demands energy. Mechanochemistry should be considered in the context of green chemistry. Green chemistry is best defined as the utilization of a set of principles that reduces or eliminates the use or generation of hazardous substances in the design, manufacture, and applications of chemical products (Anastas and Warner 1998). The development of selective and efficient synthetic methods is one of the major goals in organic chemistry. On the other hand, these new procedures should also be compatible with our environment and preserve our resources.

7.5 REGULARITIES IN MECHANICAL ACTIVATION OF ORGANIC REACTIONS

In ball milling conditions, the heavier ball materials are, the more input of mechanical energy into the reacting system takes place. For the reactions between organic and inorganic participants, the higher plasticity of the organic reactant and the lower the defect concentration in an inorganic reactant lattice lead to the lower the level of mechanical activation. Organic substances are more brittle than inorganic ones. The brittleness favors transformation of mechanical into chemical energy. Softness and fusibility of an organic sold predetermine the transformation of mechanical energy in activation energy of melting.

As applied to conditions of boundary lubrication, organic additives serve as the source of carbon. Steel carbidization enhances its wear resistance. Formation of globules in which inorganic products of additive destruction are confined in a polymeric (oligomeric) shell leads to improved antiwear and load-carrying abilities of lubricants.

7.6 ORGANIC MECHANOCHEMISTRY AND BIOENGINEERING

By virtue of our modern lifestyle, more than one third of people will likely experience the failure of a native hip joint sooner or later. Fortunately, hip joint arthroplasty is on hand to deal with this obstacle. A major unsolved problem, however, is the mechanical wear of the artificial joint, which seriously limits the joint lifetime. Undoubtedly, this also is a problem of organic mechanochemistry. Although most lubrication of the healthy natural joint relies on a film of synovial fluid, the artificial joint consists of synthetic materials that are mainly lubricated in boundary regimes. At present, in a widely used arthroplastic design, the acetabular cup consists of linear plastic ultrahigh molecular weight polyethylene, and the femoral head is a polished metal or ceramic ball. The wear of the polyethylene lining is generally regarded as the key lifetime-limiting factor. The ceramic balls are hydrophilic, and therefore endogenic proteins are adsorbed onto them without denaturation. The results are biocompatibility and better lubrication within the implant. In artificial joints, proteins decompose and form interconnected carbon-based sheets similar to graphite that prevent wear (Wimmer et al. 2003). The current implant technology works, but is still plagued by premature wear. One of the reasons is that the components involved in metal-to-metal, metal-to-ceramic, and ceramic-to-ceramic contacts are not present in the natural system. The urgent target is to develop new parts that can emulate the natural cartilage and its components in the body. Initial studies of the polymer poly(L-lysine)-*graft*-poly(ethylene glycol) demonstrated promising such emulating antiwear performance (Canter 2004 with reference to works by Drs. Perry at the University of Houston and Nicholas Spencer at ETH-Zurich).

7.7 EXAMPLES OF INNOVATIONS AT THE BORDER OF ORGANIC MECHANOCHEMISTRY

Wax compositions for skis and snowboards were proposed comprised of fullerenes or their derivatives, paraffin waxes with a melting point between 45 and 120°C, and α,γ-dibutylamide of *N*-lauroyl-*L*-glutamic acid (as a viscosity controller). The mixture was melted, stirred at 70°C for 6 h, and solidified in a mold to give a wax showing good lubricating properties (Aoyama and Suzuki 2005).

A patent was claimed to manufacture shoes with flexural fatigue resistance. The shoes contain soles or coverings prepared from leather substitutes filled with 0.1% fullerenes. Not only flexural but also the wear resistance of the shoes were excellent (Naka et al. 2005).

Cellulose molecular modifications find diverse applications. Many efforts were directed toward perfecting present techniques. Endo and Ago (2004) proposed a mechanochemical method for modification of cellulose. By milling native cellulose with toluene, the cellulose particles with a flakelike shape are obtained. The particles are formed through the exfoliation of cellulose microfibrils during milling with toluene. With water instead of toluene, crystal transformation from cellulose I to cellulose II proceeds. Cellulose molecules are most flexible when the water content is 30%. Such a condition is favorable for molecular rearrangement under external mechanical forces.

Mechanochemical technology for preparing starch gel materials was developed and applied at the Russian textile manufacturers. This technology enables reduction of the costs of textile sizing and printing processes. The concentrated water–starch suspension is mechanically activated in a rotor-impulse device built in the corresponding supply line. In doing so, the mechanical treatment is combined with pumping. The activation consists of high shear stress, vibrations and turbulent pulsations, and cavitations. The initial suspension is a dispersion of solid spherelike particles. The particles are nonswollen starch grains. Mechanical activation results in extraction of the surface lipids, cavitational erosion of the grain surfaces, combined with changes in the starch hydration degree. In alkaline media, the starch paste is formed with no heating. This excludes the necessity for steam to heat the mass. As compared to the conventional starch thickening, the mechanical treatment brings significant savings in consumption of energy and of starch itself, for the latter up to 30% (Lipatova 2001).

Organic mechanochemistry, as it follows from this book, covers organic reactions initiated mechanically. There are also organic reactions initiated by shock waves or sonication in gases and liquids. These branches of organic chemistry progressed independently and have been generalized in special books. The corresponding references were mentioned throughout the chapters here. This book is aimed to fill the gap existing for organic mechanochemistry. It is an attractive target to trace interconnections among all kinds of mechanically activated organic reactions — of course, at the 21st century level. Such a task should be considered a motion toward Wilhelm Ostwald (1909 Nobel prize in chemistry). In 1919, Ostwald proposed to distinguish special chemistry subdivision concerning all mechanical effects, diverse in their origins, on various chemical processes, irrespective of the aggregate state of the reaction participants.

Modern time, of course, introduced its own accents in Ostwald's definition. When applicable, the present book, addresses problems relating to ecology, biomedicine, mechanical engineering, and so on. I hope that concepts of this book may generate new questions and help obtain answers needed in everyday practice, and that we all benefit by development of this science and gain stimuli for our future activities. The main idea of this book can be put into the following words: science is a reasonable and worthwhile affair, but not a promenade alley for the elite — the elite that craves to manifest itself.

REFERENCES

Anastas, P., Warner, J.C. *Green Chemistry: Theory and Practice* (Oxford University Press, New York, 1998).

Aoyama, Yo., Suzuki, Yu. (2005) *Jpn. Pat.* 2005132943.

Canter, N. (2004) *Tribol. Lubr. Technol.* **60**, 43.

Endo, T., Ago, M. (2004) *Cellulose Commun.* **11**, 74.

Furey, M.J., Kajdas, Cz., Kempinski, R. (2002) *Lubric. Sci.* **15**, 73.

Lipatova, I.M. (2001) *Tekstil'naya Khim.*, **1**, 72.

Naka, Y., Miyauchi, A., Tanaka, W. (2005) *Jpn. Pat.* 2005046605.

Ostwald, W. *Handbuch der allgemeine Chemie* (Ambrosius Barth, Leipzig, Germany, 1919, Band **1**, 70).

Wimmer, M.A., Sprecher, C., Hauert, R., Taeger, G., Fisher, A. (2003) *Wear* **255**, 1007.

Zink, J.I., Hardy, G.E., Sutton, J.E. (1976) *J. Phys. Chem.* **80**, 248.

Author Index

155

Subject Index

A

Acid-base coordination, 85, 86, 90, 92
Acylation, 73, 74
Adamantane derivatives, 41
Additives
    antagonism, 50–52
    synergism, 50–52
Amonton's law, 31

B

Bioavailability, 7, 86, 91
Biocompatibility, 125, 128, 132, 134
Biodegradability, 9, 47, 48, 125

C

Carbidizing, 42, 43, 49, 139
Catalysis, 72, 79–82
Ceramics, 46, 79, 140
Charge-transfer complexes, 16, 17, 85, 86
Cholesterol esters, 107, 127
Citronellol, 68, 69
Coloration, 11, 19–23, 26, 27
Condensation, 46, 69, 70, 80
Conformation, 16–26, 85, 90, 95–97
Coordination
    to metals, 75–79, 93, 94, 130
Crack
    reaction zone, 4–8, 22, 23, 129
Cyclization, 80, 81
Cycloaddition, 70, 83, 84
Cyclodextrins, 21, 86, 87, 91

D

Damage sensors, 12, 19, 25, 27
Dehalogenation, 74
Deswelling, 112–115
Devulcanization, 66
Diffusion, 4, 7, 41, 113
Dispersions
    solid, 86, 91
Doping, 32, 41

E

Ecological aspects, 46–48, 56, 57, 75, 80, 138, 139
Economical aspects, 57, 75, 109, 138–142
Electron transfer, 4, 6, 21, 83, 84, 92–96, 106
Esterification, 40, 47, 72, 73
Exoelectrons, 2, 5, 19–22, 34–36, 49, 93
Extrusion, 24, 50, 62, 111

F

FCK rule, 105
Fluoropolyalkylethers, 34, 35, 37–39, 113
Friedel–Creagh–Kmetz rule, 105
Fullerenes, 11, 54, 55, 82–84, 125–128, 140
Fulleroids, 126
Fullerols, 84, 87
Functionalization, 84

G

Gels, 112–115, 131

H

Homolysis, 2, 5, 6, 21, 22, 52, 53, 64–66
Host-guest coordination, 18, 72, 77, 85–88, 107, 127
Hydrogen-bond complexes, 88–92, 97
Hydrogen bonding, 4, 7, 16, 26, 38, 39, 72
    between biomolecules, 114, 115, 121
Hydrolysis, 43

I

Indandiones, 2, 3, 5, 6, 21, 22
Isomerization, 23, 26

J

Jumping
    electron, 95

K

Kramer's effect, 19, 32, 62, 63, 93, 106, 117, 118